Metal
Matrix
Composites

Authors:
C. T. Lynch
J. P. Kershaw
Air Force Materials Laboratory
Wright-Patterson AFB, Ohio

CRC Press
Taylor & Francis Group
Boca Raton London New York

CRC Press is an imprint of the
Taylor & Francis Group, an **informa** business

CRC UNISCIENCE SERIES

Another information program recently initiated by CRC is the "Uniscience" Series. The function of "Uniscience" has been defined as the publication of highly authoritative scientific reference works dealing with those subjects for which there exists an urgent necessity for published information due to significant new developments.

To achieve this function, the "Uniscience" program utilizes the conventional reference book format, but with one, all-important difference: it has radically reduced the traditional lag between completion of the manuscript and publication of the printed book. Utilizing newly developed editorial procedures together with newly perfected computerized type-setting equipment, the time-span normally required from manuscript to printed book has been dramatically shortened to a matter of only a few months.

This very real "break-through" in production time makes it possible for the Uniscience Series to provide a new dimension in the distribution of scientific information.

B. J. Starkoff, President
THE CHEMICAL RUBBER CO.

RADIOLOGICAL SCIENCES
  Yen Wang. M.D., D.Sc. (Med.)
  University of Pittsburgh

SPECTROSCOPY
  Jeanette Grasselli, M.S.
  Standard Oil Company (Ohio)

  W. M. Ritchey, M.S.
  Case Western Reserve University
  James W. Robinson, Ph.D.
  Louisiana State University

TOXICOLOGY
  Irving Sunshine, Ph.D.
  Cuyahoga County Coroner's Office, Ohio

CRITICAL REVIEW JOURNALS

ANALYTICAL CHEMISTRY
  Louis Meites, Ph.D.
  Clarkson College of Technology

BIOCHEMISTRY
  Gerald Fasman, Ph.D.
  Brandeis University

BIOENGINEERING
  David G. Fleming, Ph.D.
  Case Western Reserve University

CLINICAL SCIENCES
  Willard R. Faulkner, Ph.D.
  Vanderbilt University Medical Center
  John W. King, M.D., Ph.D.
  Cleveland Clinic Foundation

ENVIRONMENTAL SCIENCES
  Richard G. Bond, M.S., M.P.H.
  University of Minnesota
  Conrad P. Straub, Ph.D.
  University of Minnesota

FOOD AND NUTRITION
  Thomas E. Furia
  CIBA-GEIGY Corp.

MACROMOLECULAR SCIENCE
  Eric Baer, Ph.D.
  Case Western Reserve University
  Phillip Geil, Ph.D.
  Case Western Reserve University
  Jack Koenig, Ph.D.
  Case Western Reserve University

MICROBIOLOGY
  Allen I. Laskin, Ph.D.
  Esso Research and Engineering Co.
  Hubert Lechevalier, Ph.D.
  Rutgers University

RADIOLOGICAL SCIENCES
  Yen Wang, M.D., D.Sc. (Med.)
  University of Pittsburgh

SOLID STATE SCIENCES
  Richard W. Hoffman, Ph.D.
  Case Western Reserve University
  Donald E. Schuele, Ph.D.
  Bell Telephone Laboratories

TOXICOLOGY
  Leon Golberg, D.Phil., D.Sc.
  Albany Medical College of
  Union University

TABLE OF CONTENTS

FOREWORD

The inception of this book dates back several years to the suggestion and encouragement of Professor Lynn Ebert of Case Western Reserve University that I prepare a review of the mechanical properties of metal matrix composites for the CRC Critical Review Series. The major objective of that review was to make a critical assessment of the state-of-the-art, probing the problem areas as well as achievements, and recommending promising avenues for further efforts. I was privileged to have the assistance of John Kershaw and Brent Collins in that endeavor. Our approach was to place emphasis on those metal matrix systems or combinations which were most highly developed as potential engineering materials, rather than try to give an exhaustive review of the field. The reception of that effort was indeed heartening, and sufficient to lead to a more comprehensive treatment of metal matrix composites.

In this book we have retained the emphasis on those systems which have received the most attention in research and development. Thus, the mechanical properties sections are predominantly on the properties of boron/aluminum and boron/titanium composite systems, which are the best characterized metal matrix composites at the present time. This predilection will be noted in the treatment of other subjects such as consolidation, fabrication, compatibility, and applications. There are other sections which necessarily examine systems which do not have this pretext of a substantial background of property data. These include the discussions of refractory metal wire composites and eutectic composites. These systems show considerable potential for the higher temperature regimes in particular. For the sake of completeness, a reasonably detailed treatment of whisker reinforced metals has been included, although the presumption is made that continuous filament reinforcement will continue to dominate research interest and follow-up development programs.

We have attempted to give an adequate coverage of the field of metal matrix composites within the stipulated boundaries, and to retain a critical flavor to the presentation. Thus, we have included evaluation and assessment in addition to objective accounting of scientific investigations. In exercising this prerogative, no attempt has been made to cover all of the available literature. This may well be an emerging field of materials science in its infancy, but the volume of recent published work is surprising. So our apologies to anyone whose work has been omitted because of space limitations, or the specific emphasis of this treatment.

A special section on the latest reported mechanical properties of boron/aluminum and boron/titanium composites was added to the manuscript just before going to press. In this manner, we hope to have achieved more timeliness than generally realized in a book.

Our grateful thanks to our many colleagues in industry, academic circles, and in government for their assistance in supplying information on their work, and their commentary on the significance of the many diverse efforts underway. My thanks also to John Kershaw for associating with me on this venture, and to the Editorial Staff and Jerry Becker of the Chemical Rubber Company for their assistance.

I want to pay special tribute to my wife, Betty Ann, for her wonderful patience, encouragement, and help and to our children, Karen, Ted Jr., Richard, and Thomas, for giving up some time with Dad while he wrote. Finally, I want to thank two excellent secretaries, Jean A. Gwinn and Kayann Pickens, for their superb aid in the preparation of this book.

C. T. Lynch
Fairborn, Ohio

THE AUTHORS

Charles T. Lynch, Ph. D., is Chief, Advanced Metallurgical Studies Branch, Air Force Materials Laboratory, and Senior Scientist.

Dr. Lynch was graduated from The George Washington University, with a B.S. (Chemistry) degree in 1955. He received the M.S. and Ph.D. (Analytical Chemistry) degrees in 1957 and 1960, respectively, from the University of Illinois, Urbana.

In 1960 Dr. Lynch came to the AFML as a Project Officer, serving on active duty until 1962. He served successively as a research materials engineer, group leader, and lead scientist for in-house research of the Ceramics & Graphite Branch from 1962-66.

Dr. Lynch has won numerous awards and is listed in *Who's Who in the Midwest Dictionary of International Biography* and in *American Men of Science.* He has published more than 60 research papers and holds 12 patents.

John P. Kershaw, Sc.D., is Assistant Chief, Metals and Ceramics Divison, Air Force Materials Laboratory. He is currently on assignment at the Air Command and Staff College, Air University, Maxwell Air Force Base, Alabama.

Colonel Kershaw was graduated from Syracuse University in 1957 with a B.Ch.E. (Chemical Engineering) degree and received a Sc.D. (Materials Engineering) degree from Massachusetts Institute of Technology in 1964.

In 1958 Colonel Kershaw entered the Air Force as a Lieutenant in Munitions and Disaster Control. He then served in a series of research positions and as an Associate Professor in the Department of Engineering Mechanics at the Air Force Academy from 1964-68. In 1968 he came to the AFML as the Technical Area Manager for Metal Matrix Composites.

Colone Kershaw is active in a number of scientific societies, particularly the AIME, American Society for Engineering Education, and American Society for Metals.

Chapter 1

INTRODUCTION

The concept of reinforcing a material by the use of a fiber is not a new one. The Egyptian brick layer employed the same principle more than three thousand years ago when straw was incorporated into the bricks. More recent examples of fiber-reinforced composites are steel-reinforced concrete, nylon and rayon cord-reinforced tires, and fiberglass-reinforced plastics. In the last several years considerable progress has been made on new composite structures particularly utilizing boron (on tungsten substrate) fibers in various matrices. Many of these advances have been reviewed recently by P. M. Sinclair[1] and Alexander, Shaver, and Withers.[2] An excellent earlier survey is available by Rauch, Sutton, and McCreight.[3] Boron-reinforced epoxy composites are being fabricated and tested as jet engine components, fuselage components, and even as a complete aircraft wing because of the tremendous gain in experimentally demonstrated properties such as modulus, strength, and fatigue resistance, particularly on a weight normalized (e.g., strength/density) basis. Other than glass/epoxy and boron/epoxy composites and perhaps boron/aluminum, the systems now under study are in the early stages of research and development. These include other boron/metal composites, graphite/polymer, graphite/metal, graphite/graphite, alumina/metal, and aligned eutectic (directionally solidified) combinations. As Sinclair points out, designers are wary about filamentary composites because "there is little background information and scant experience."

The paucity of mechanical data, difficulties in fabrication, high costs, and degradation reactions at anticipated use temperatures have all contributed to the cautious development of reinforced metals. There is also a natural reluctance of the designer to undertake new design concept development. This is required since conventional concepts do not optimize the use of fiber composites. Competitive advantages that metal matrix composites offer over conventional metallic systems include the ability to: (1) take advantage of the anisotropic character of the composite in the efficient design and fabrication of structures,

(2) tailor-make a material to meet a specific set of engineering strength or stiffness requirements, and (3) increase the stiffness, strength, and thermal stability of such common engineering alloys as aluminum, titanium, and nickel. It is not unusual to demonstrate weight savings to 40% on specified structures, often with significant increases in fatigue strength, stress rupture, and creep properties.

In comparison to other composite matrices, the most obvious potential advantage of the metal matrix is its resistance to severe environments, toughness, and retention of strength at high temperatures. Burte and Lynch[4] have emphasized that the most unambiguous potential for future use is at temperatures above 1600° to 1800°F (871° to 982°C) when considering the competition from the more familiar directions of metallurgical development or polymer matrix composites. In a composite structure it is possible to emphasize environmental stability of the matrix at elevated temperatures since the required mechanical strength and stiffness can be obtained from the reinforcement. In homogeneous materials, mechanical properties and environmental stability may have to be compromised to produce a useful structure. In filamentary reinforced metals the shear strength requirements of the matrix are nominal since the matrix serves only to transfer load into the filaments. Figure 1[5] schematically illustrates the broad range of temperatures over which composite structures with various matrix materials may be considered. Up to about 1000°F (538°C), several potential metal composite systems have already been identified. For high temperatures, where oxidation presents a major problem in the development of ductile structural materials, stable reinforcements such as continuous alumina filaments have considerable potential. At temperatures below 600°F (315°C) the potential is less clear; but possible advantages of a metal matrix compared to a polymer matrix include improved abrasion and erosion resistance, greater shear, bearing, and transverse strengths, increased toughness, the ability to use conventional metal processes, and greater deformation

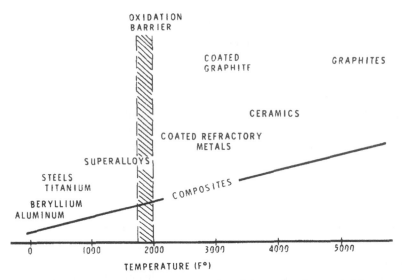

FIGURE 1. Temperature regimes of usefulness of various materials.

prior to fracture. Disadvantages include higher density, more difficult fabrication, matrix-filament reactions, and higher cost. The metal matrix can provide a meaningful contribution to the strength and elastic modulus in all directions.

A body of important mechanical property data has been obtained within the last two years. This review concentrates on the recent literature, partly because this is where most of the information on limiting mechanisms and properties is found, and partly because much of the earlier work was conducted on inadequately characterized materials, sometimes poorly bonded or containing reinforcing filamentary materials with widely disparate mechanical properties. In many instances, as Tsai suggests,[6] the unique character of filament-reinforced metals demands more sophisticated mechanical tests than have generally been applied. The anisotropic character of composite structures demands substantial emphasis on properties associated with multiaxial stress states as opposed to the early reliance on uniaxial tensile data to characterize these materials. With an increasing availability of a variety of high strength, high modulus continuous filaments, the number of potential filament/metal systems is rapidly increasing. Advances in fabrication techniques have produced composite

structures with reproducible properties, the most advanced being boron/aluminum.

Emphasis in this review has been placed on systems where a considerable body of mechanical data has been obtained and can be critically examined. It is those systems which are closest to engineering development that the authors felt would be of most interest to the reader. Of necessity, several areas were not covered in much detail, for instance, graphite-reinforced metals, where less effort has been expended to date and the reader is referred to the cited references for further information. A separate section was included on whisker-reinforced metals as unique composite structures; but generally, these materials have not gained the prominence predicted for them ten years ago. Directionally solidified eutectics, which are sometimes considered a special case of whisker-reinforced structures, have also been considered separately. A section on mechanics has been included which suggests the available methods to predict mechanical properties of composite structures and the response to various external loads. It is mainly a compilation of referenced approaches since little has been done to apply more than a simple "rule of mixtures" analysis in most metal composites. It appears to the authors that this is an area where much could be accomplished.

Chapter 2

REINFORCEMENTS

During a period of less than ten years a wide selection of high modulus, high strength, low density, and often refractory filamentary materials has become available as candidate reinforcements for metallic materials. A number of these are listed with typical mechanical properties in Table 1, with the more familiar glass and metal wire filaments given for comparison. For metals the filaments of major interest have been boron, silicon carbide, alumina, refractory wires, and graphite. In many cases the preparative methods result in problems such as defects and residual stresses, which mitigate against maximizing the mechanical properties of composite structures. A silicon-carbide coated boron with slightly lower tensile strength than boron (Borsic®) is also made and used where boron-metal interactions degrade the filament.

Boron filaments are produced by chemical vapor deposition of boron on a hot 0.5 mil tungsten wire substrate[7] from boron trichloride and hydrogen at approximately 1000°C. The typical filament is 4 mils in diameter with strengths averaging 450 x 10³ psi. A filamentary fatigue life in excess of a million cycles (using a tension-zero-tension cycle at 150 cycles/minute) has been measured using a cyclic load of half the mean tensile strength. The density of 2.6 g/cm³ is slightly greater than E-Glass, but the specific modulus of boron is far superior to that of glass fiber.[8,9]

The boron deposit is extremely fine-grained, bordering on being amorphous.[10] The surface typically has a "corn-cob" appearance (unless contaminated with oil or grease films which are sometimes found on "as received" filaments).[11] During the deposition process the tungsten wire is at least partially converted to tungsten diboride[12] and cooled rapidly from the hot zone in high-speed processing. Residual stresses are generated in the boron filaments from the differences in thermal expansion between the boron deposit and substrate. These residual stresses make the filaments susceptible to longitudinal cracking, as shown in Figure 2. Complete fiber-splitting in 1-ft lengths of boron filaments was observed by Burte, Bonanno, and Herzog.[11] This fiber-splitting is apparently responsible for low transverse strengths

TABLE 1

Filament Properties

Material	Tensile Strength (ksi)	Elastic Modulus (X 10⁶ psi)	Density (g/cm³)	Density (lb/in.²)	Strength/Density Ratio (X 10⁶ in.)
Beryllium	180	45	1.84	0.066	2.73
Molybdenum	320	48	10.2	0.369	0.88
Steel	600	29	7.9	0.286	2.10
Tungsten	575	59	19.3	0.697	0.83
E-Glass	500	10.5	2.5	0.092	5.43
S-Glass	600	12.5	2.5	0.092	6.52
Silica	500	10.5	2.2	0.079	5.43
Boron (on Tungsten)	450	58	2.6	0.096	4.68
Graphite*	285	50	1.66	0.060	4.75
Silicon Carbide (on Tungsten)	350	55	3.4	0.123	2.84
Alumina	350	70	3.98	0.144	2.43
Alumina Whiskers	2000	70	3.98	0.144	13.9
Silicon Carbide Whiskers	1500	70	3.2	0.115	13.0

* There are many types varying in strength (150–375 ksi), modulus (25–75 X 10⁶ psi), and density (1.5–2.0 g/cm³).

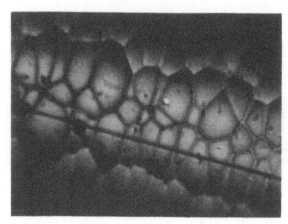

FIGURE 2. Surface of boron filament with longitudinal crack (2000x).

in unidirectional boron/aluminum conposites which will be examined in more detail in the mechanical property discussions.

Substrates other than tungsten have been examined to reduce the cost and also produce a boron filament having a density somewhat closer to that of boron (2.34 g/cm^3). Fused silica is one example. Diborane has been decomposed at 700°C to form boron on silica, the lower temperature of deposition necessary for the substitution of the silica substrate for tungsten.[8,11] The silica-boron filament had strengths up to 350,000 psi and a density of 2.35 g/cm^3, but the deposition process was difficult and it has not replaced tungsten-boron. To retard degradative reactions in metals, more inert materials such as silicon carbide[14] and boron nitride[15] have been applied in thicknesses of less than 1 mil.

Silicon carbide has been deposited directly on tungsten[16] using ethyltrichlorosilane (CH_3SiCl_3), hydrogen, and argon at approximately 1300°C This is a difficult synthesis by vapor deposition and the filament properties have varied considerably. Silicon carbide has a higher density at 3.4 g/cm^3 but the retention of strength at higher temperatures is very good. Recent strength values have exceeded 350,000 psi.

Single crystal alumina filaments have been produced from molten alumina utilizing a carefully controlled drawing process.[17] The continuous filament is typically about 10 mils in diameter, with a strength of 350,000 psi and with high residual stresses as-grown. Voids within the crystals and morphological defects have been found which presently limit the filament strength.

The strength has been found to be extremely sensitive to the presence of surface defects.[18] The density is comparatively high at 3.98 g/cm^3. These problems are offset, however, by the excellent high temperature properties of alumina and its stability in many transition metals of interest.

Considerable effort has been expended in recent years to produce high strength graphite fibers. Fiber synthesis usually involves pyrolysis and high temperature processing of polymeric precursor materials, such as polyacrylonitrile or rayon filaments which are graphitized at temperatures >2600°C.[19] Graphite yarn is available which exhibits relatively high strengths and moduli. Densities range from 1.4 to 1.8 g/cm^3 with higher density materials having higher strengths and moduli. For metals, the yarn, with individual fibers well under 10μ in size, presents a formidable handling problem. This, coupled with mechanical and chemical incompatibility problems of graphite with most metals, limits their current potential. Jackson,[20] for example, has found that at relatively low temperatures (600°C) contact with metals such as aluminum, nickel, cobalt, copper, and platinum causes some structural recrystallization of the surface of graphite filaments, causing a drastic loss in strength. Thus, the excellent retention of strength by graphite filaments at high temperatures is of no real advantage in most metal matrices. Despite these observations, much disappointing work continues on fine-diameter graphite yarn. It is quite possible that with the development of a large diameter monofilament of 2 to 4 mils' size, the surface recrystallization reaction will have a tolerably small effect on filament strength. The other problems of solubility, carbide formation, and thermal expansion mismatch will then have to be solved.

Figure 3 gives the short-term elevated temperature tensile strengths of a number of candidate reinforcements. These were obtained from several sources[1,3,11,21-23] and from work done in the Advanced Metallurgical Studies Branch, AFML. Since the time at temperature differs, as does the time required to reach test-temperature in various apparati, the results serve mainly as a qualitative comparison. It may be obvious, but should be noted nonetheless, that data on short sections in flexure are often significantly higher than tensile values on long specimens of the same filaments. Thus, some confusion is found when inordinately

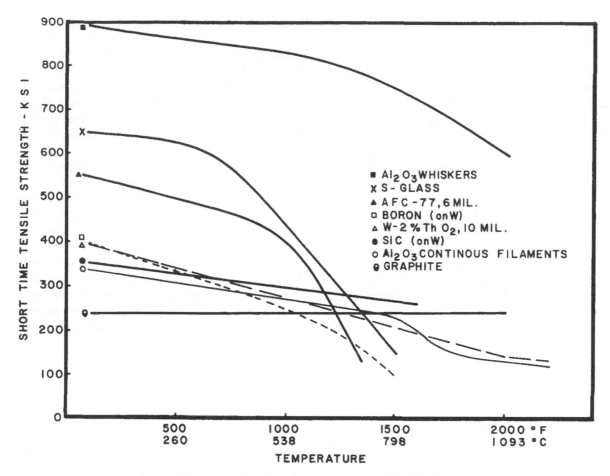

FIGURE 3. Short-time elevated temperature strength of reinforcements.

high strength values are reported without clarification.

In Figure 3 the strengths of some filaments, such as alumina and refractory metal wires, are seen to be stable over a broad temperature range. Others, such as high strength stainless steel, are excellent at room temperature but fall rapidly with increasing temperature. For SiC the data are estimated from previous results[11] and the increased strength that has been obtained recently for this filament. Sinclair[1] gives slightly higher values for SiC filament strengths above 500°C.

In terms of improving the stress rupture and creep resistance of niobium and nickel, there seems to be some potential for refractory wire reinforcement. Petrasek and Signorelli[23] have recently evaluated tungsten-2% thoria (shown in Figure 3), tungsten-2%-5% rhenium, 218 CS tungsten (lamp filament wire), and tungsten-1% thoria reinforcements in nickel. Although some interesting properties have been obtained, serious

reactivity problems and consequent fiber degradation are often observed.[24] These problems have been reviewed in the subsequent section on refractory metal wire reinforced composites. The poor specific strengths and specific moduli of these refractory metal wires, as seen in Table 1, mitigate against any applications other than at very high temperatures (2000°F – 1093°C).

An excellent review of the methods of preparation and properties of various reinforcement materials is found in a book by Galasso.[24] The large amount of work on boron filaments, for example, is well documented and referenced. High temperature properties, reactions such as oxidation, and coatings research for filament protection in active matrices at higher temperatures were considered by Galasso. The reader is referred to this reference for more detailed information on the many filamentary materials that have been investigated in the last decade. In recent years improvements in processing conditions have

5

eliminated much of the property variance in high volume produced filaments such as boron. Some of the problems such as growth nodules, surface damage, surface contamination, cracks and voids formed during the growth process, etc. recur with each new filamentary material and cause much difficulty in subsequent processing research and mechanical property correlations. The cracks sometimes observed in boron filaments are illustrated in Figure 2. Close examination of this type of axial crack indicates that it is formed during the growth process.

In addition to defects which lower the filament strength, the condition of the surface is important, particularly as it may affect the bonding conditions and interface in a metal matrix composite or the character of the tape in a preform material. Typical surface contamination which has been observed includes oily films and miscellaneous dirt and debris. These types of impurity contaminants are illustrated in Figure 4. In Figure 4A the surface has an oil film, and in Figure 4B the surface contains dirt and debris. In metallurgical laboratories, often the debris has been found to contain extraneous metal filings. The filament can be cleaned in a solvent, brushed, or sometimes heat treated at relatively low temperatures to remove the surface contaminants. Careful handling and storage can avoid much of this problem.

In Figure 3 the short-time elevated temperature tensile strength of boron filament is shown to drop gradually up to temperatures exceeding 500°C. Beyond this point it begins to drop rapidly. Galasso has reviewed some of the high temperature investigations of boron filament and found that the room temperature strength after heat treatment in argon showed little degradation to a temperature of 700°C.

On the other hand, the room temperature strength dropped rapidly after air exposure at temperatures above 500°C. This corresponds to the point where substantial oxidation and consequent fiber degradation occur. Long-time air exposure at lower temperatures also seriously degrades the boron filament strength. After 1,000 hours at 200°C, for example, a 40% loss in strength was observed. This contrasts to the stability of the silicon carbide coated boron, where no loss in strength occurs after 1,000 hours in air at temperatures up to 600°C. The protection against oxidation extends to the stability of the silicon carbide coated boron after treatment in a 2024 Al alloy. After 1,000 hours at 600°C the fiber strength is 70% of the virgin strength, contrasted to about 35% for uncoated boron after only 100 hours at 600°C.[24] A boron filament with a silicon carbide film is shown in Figure 5. The silicon carbide coating is very thin, comprising about a 0.1 mil thickness on a typical 4-mil boron filament. It is an even coating with excellent adherence to the substrate boron. An example of an incomplete coating is shown in Figure 6, where nickel has been deposited on boron filament.[33]

FIGURE 5.   Boron filament with surface silicon carbide film (1000x).

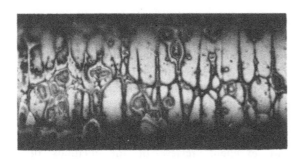

FIGURE 4B.   Dirt and debris on boron filament surface (800x).

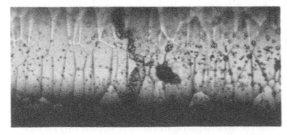

FIGURE 4A.   Oil film on boron filament surface (800x).

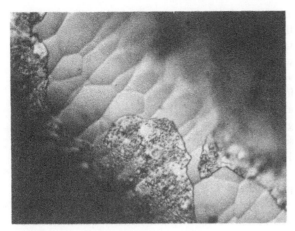

FIGURE 6. Nickel-coated boron filament with discontinuities in the film (1000x).

The underlying corncob structure of the boron is very distinct.

In one series of experiments[25] boron filaments were subjected to short-time heat treatments in cycles between room temperature and 600°F (315°C). The results are shown in Figure 7 as a normalized strength compared to the untreated filament strength (σ/σ_o) as a function of the number of cycles. Initially, the strength increases, possibly due to annealing of surface flaws, and then the strength drops rapidly as the number of cycles is increased. The structure is definitely affected by this treatment, as illustrated in Figure 8 for filaments cycled 10,000 times. In some places the corncob structure is almost obliterated or heavily marked by some type of surface attack or change while in other areas the fibers seem to

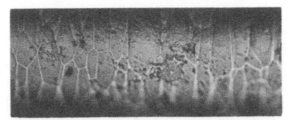

FIGURE 8A.

FIGURE 8B.

FIGURE 8. Boron filaments cycled 10,000 times to 600°F (315°C) (800x).

have changed very little. For specimens treated for 1,000 cycles, the effects are much less pronounced, and for 100 cycles or less, not observable. It may be that the strength degradation comes from differences in thermal behavior of the borided tungsten core and the boron outer portion of the fiber, rather than from the surface changes observed in this study.

Many new filamentary materials have appeared from laboratories around the world, often with rather disappointing properties. From some will come, however, the high strength, high modulus reinforcements for the next generation of composites. A few of these experimental filaments are presented here to indicate the wide variety of possibilities with some of the problems that are evident in their synthesis.

Kaolin fibers have been made which are transparent, high density filaments with smooth surfaces. In Figure 9A a regular cross section portion of a kaolin fiber is illustrated. Figure 9B illustrates the growth nodules which were found irregularly but often on the fiber lengths. Some structures which appeared to be voids elongated in the fiber axis direction were also observed. The modulus values were rather consistent at an average of 15 to 16 x 10^6 psi, with a high value of 22 x 10 psi. The strength values showed considerable scatter, with average values of slightly over 100 ksi, except for one fiber which had a high value of 239 ksi. Cross section diameters varied considerably from 2 to 5

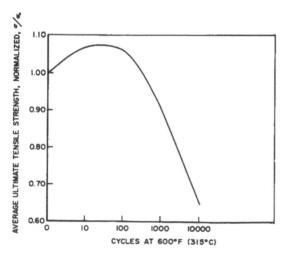

FIGURE 7. Boron filament strength as a function of cyclic heat treatment.

7

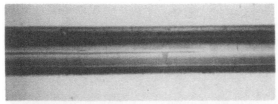

FIGURE 9A.

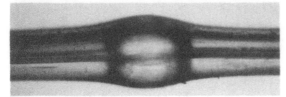

FIGURE 9B.

FIGURE 9. Kaolin fibers (104x). A. Smooth surface. B. Irregular surface.

mils. It is quite possible that a fiber like this would be attractive if the properties hold up at moderately high temperatures and can be optimized for the process.

One of the newer filaments that we have examined is a smooth glassy carbon filament. Actually, the striations seen on one of these filaments in Figure 10A are very pronounced at higher magnifications and the filament has a somewhat granular appearance. The twist seen here has observed in many of these filaments. It is more pronounced in the filament shown in Figure 10B, which also illustrates a growth nodule

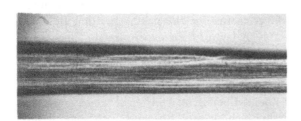

FIGURE 10A. Thick carbon fiber showing smooth surface and a twist (280x).

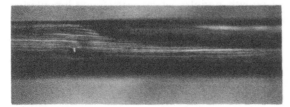

FIGURE 10B. Thick carbon fiber showing twist and growth nodule (280x).

and a markedly larger cross section. The carbon filaments had diameters which varied from less than 0.5 mil to 2 mils. Ultimate tensile strengths were disappointing at 70 to 100 ksi, with some values below 40 ksi. The elastic modulus was even lower, at 4 to 6 x 10^6 psi, with a few values around 10 x 10^6. Considering that commercial graphite yarn-type filamentary reinforcements such as Thornel have elastic moduli of 25 to 75 x 10^6 psi, these values are rather poor.

Beryllium has always elicited great interest as a possible reinforcement with its high modulus and high strength to density ratio, as seen in Table 1. Cost, brittle behavior, and toxicity questions have always been drawbacks. Beyond this, most of the beryllium fibers have had poor properties. (This should be distinguished from the larger diameter beryllium wire available in limited fashion commercially, which has been used in some metal matrix composites research. The big drawback to the larger filament has been an excessive cost of $5,000 to $10,000 per pound of filament.) Short beryllium fibers of 1 to 2 mil diameter were tested and exhibited elastic moduli of 6 to 18 x 10^6 psi and ultimate tensile strengths of 16 to 28 ksi. The fibers were soft and tearing was observed upon failure. The fibers are shown at low magnification in Figure 11. The rough tearing on fracture of a beryllium wire is shown in Figure 12. The cross section of these fibers was extremely irregular. Even over short lengths of fibers this variation results in questionable mechanical property measurements. The total cross sections corresponded to a diameter of 1 to 2 mils but actually none of the fibers had a cross section resembling a

FIGURE 11. Beryllium fibers (4 1/2x).

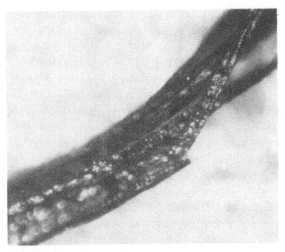

FIGURE 12.   Bend and fracture in beryllium wire (70x).

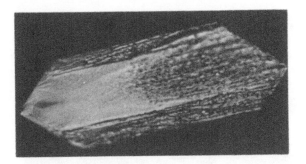

FIGURE 14.   Fracture cross section of ZrO$_2$ fiber (1070x).

circular area. Most of the filaments were extremely irregular, tending to be long, narrow, and with pointed edges.

Zirconia filaments ranging from 1 to 2 mils in diameter have been found to have very pronounced defects or growth steps. This is observed in Figure 13 at a low and high magnification. Some growth lines were also found in the longitudinal axial direction. The cross section for a zirconia fiber was very distinctive, as seen in Figure 14. It might be well to mention here that most commercial filaments have essentially circular cross sections. These include the boron,

FIGURE 13A.   ZrO$_2$ fiber showing growth bands (200x).

FIGURE 13B.   ZrO$_2$ fiber showing growth bands (410x).

Borsic, alumina, silicon carbide, and refractory wires, which constitute the principal reinforcements for metals. Some attention has been given to ribbon reinforcements to improve the loading geometry and transverse properties of metal composites, but the approach has not received much attention by commercial producers. Growth of a circular filament on tungsten by vapor deposition is a much simpler process than one to yield an off-axis deposit. The ZrO$_2$ fibers tested as-received in the late 1960's exhibited strengths of 65 ksi and elastic moduli of 15 to 20 x 10^6 psi. By 1970 the values obtained had increased markedly. Ultimate tensile strengths averaged to 100 ksi, with values as high as 184 ksi, and elastic moduli ranged from 20 to 40 x 10^6 psi. The pronounced growth defects probably substantially reduce strengths of some of the filaments. For temperatures exceeding 1100 to 1200°C, the filament would have to be stabilized in the cubic phase with CaO, MgO, or Y$_2$O$_3$ to avoid the monoclinic to tetragonal phase transformation.

Research on silicon carbide reinforcements is based on both continuous silicon carbide filaments on tungsten and on silicon carbide whiskers. The properties of both types have been indicated in Table 1. Attempts have also been made to produce silicon carbide filaments of large cross sections without a tungsten substrate. Most of these intermediate size filaments had diameters of 0.5 to 2 mils and exhibited complex growth patterns as shown in Figure 15. A substantial amount of branching was also observed as illustrated in Figure 16. The fibers were sometimes transparent, sometimes opaque, with colors ranging from brownish-yellow, to green, to an intense blue. The green is similar to the color of silicon carbide whiskers. Strengths close to 300 ksi and moduli of 50 x 10^6

FIGURE 15.   SiC fiber showing surface (280x).

FIGURE 16.   Branching on SiC filaments (550x).

psi were measured. If this filamentary material can be processed with larger cross sections and with fewer defects, the lack of a heavy and costly tungsten substrate would make it attractive.

Examination of as-received alumina filaments has often shown significant variation in the diameter, as seen in profile view in Figure 17. In one series of tests on a 5-mil alumina filament, for example, the diameter varied from 4 to 7 mils. In another series where the variation was not as dramatic, the values for another 5-mil filament ranged from 4.8 to 5.4 mils. This is a much greater variation than has been observed with boron filaments. It also complicates the measurement of tensile strength since the diameter must be carefully determined for each test run. This may account for some of the extremely high strengths

FIGURE 17.   Diameter variations of continuous Al_2O_3 filament (100x).

above 500,000 psi for this filament, which are occasionally reported. In recent work in the Air Force Materials Laboratory, the values obtained seldom exceed 450,000 psi for tensile strength[26] at room temperature and the average value, as seen in Table 1, is much lower. Another problem previously cited[18] is the large amount of void formation observed. This is difficult to see in the almost transparent, smooth surface, single crystal alumina filaments. An example of the sometimes intricate pattern of voids found in alumina filaments is shown in a dark field view in Figure 18.

Crane and Tressler[26] have suggested that internal pores control the filament strength by acting as stress concentrators which cause local plastic deformation. From these regions critical cracks then initiate and cause failure. While definitive proof of this concept is lacking, evidence of pore involvement has been observed. Much of the interest in alumina filament has been in potential high temperature applications. As shown in Figure 3, the elevated temperature strength of alumina is attractive compared to many other filaments, but there is a substantial drop in this strength above 800°C. This is attributed to plastic deformation of the filament at these higher temperatures.[27] Because sapphire filaments are extremely sensitive to surface damage, they are normally protectively coated for transit and handling. For use in organic matrices the filament is coated with epoxy material. This can cause carbon contamination on burn-out in metals where polystyrene is used to avoid this problem. Damage during storage, handling, and fabrication may consist of anything from surface self-abrasion by filaments rubbing against each other, to damage in the filament unwinding, to fabricate preforms or surface roughening by chemical interaction with matrix, or thermal etching of the filaments themselves. Flame polishing[28] can improve the quality of the drawn filament but is of no real advantage in later processing in terms of surface perfection. Coating with polystyrene affords protection of the filament through the initial processing steps, but is of no advantage in the subsequent heat treatments of fabrication to high integrity composite bodies where the coating has been burned off. Some recent work demonstrates that the strengths of self-abraded sapphire filament and the pristine (as-drawn or "virgin" filament without significant surface flaws) filament are essentially the same at temperatures above 1652°F (900°C). In the tem-

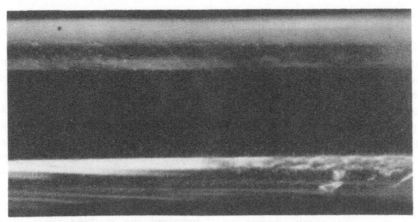

FIGURE 18.   Surface, dark field, continuous Al₂O₃ filament (300x).

perature region from 932 to 1472°F (500 to 800°C) the values obtained for the pristine filament overlap the values for the self-abraded filament.[26,27] The average strengths for the pristine filament, however, are definitely higher in this region. For a transition metal composite reinforced with alumina filament such as Al₂O₃/Ti, the protection of the filament against surface damage is of definite interest since the temperature of maximum use is determined by the overall reaction kinetics to be in the 500 to 800°C range.[29]

For possible use of alumina in more refractory composites such as Al₂O₃/Ni-Cr, consideration must be given to increasing the high temperature strength of the alumina. There are some discrepancies in the results of this type of study, apparently due to the dependence of the tensile strength values on such factors as strain rate and specimen history. Numerous studies have been conducted recently[17,26,30] which do not entirely agree on the shape of the strength-temperature curve. The key point, however, is the need to improve high temperature strength. This may be approached by reducing the number of voids present and by decreasing the propensity for plastic deformation by modifying the filament.

Current practice of drawing alumina filaments from the melt[17] has produced a reinforcement with excellent potential for further development. A few of the attempts to produce a more acceptable process for mass production at low cost should be mentioned. These approaches and major problems encountered include the following efforts:

1. Formation of polycrystalline alumina from a pure alumina melt by melt spinning. There has been extreme difficulty in this process handling the abrupt transition from the melt to the solid.

2. A similar process to the above where the fiber is sheathed in Vycor. The molten alumina and molten Vycor are forced through concentration nozzles. Again, the major problem is maintaining structural integrity of the alumina through the melt to the solid, even with the use of the Vycor.

3. Formation of polycrystalline alumina by passing a tungsten substrate through a molten alumina bath. Adherence and thermal expansion differences are major problems here.

4. Fiber spinning from an aluminum chloride salt bath with silica sols and boric acid additives. Mullite and alumina have been formed in this manner but the lengths tend to be short and properties have been disappointing.

5. Chemical vapor deposition of alumina on a tungsten substrate has been done using the hydrolysis of oxidation of aluminum chloride and the decomposition of organoaluminum compounds. Difficulties have included nonuniform deposition and contamination of the deposit.

6. Thermal conversion of precursor alumino-boro-phosphate glass filaments drawn from a high alumina content melt has been studied. Difficulties include problems in the conversion from the glass to alumina filament and low strengths.

Difficulties in obtaining suitable continuous alumina filaments can be seen in the recent work of Gruber and Hill.[31] Their major effort was to employ the water-gas reaction with aluminum chloride as follows:

$$Al_2Cl_6 + 3CO_2 + 3H_2 \rightarrow Al_2O_3 + 3CO + 6HCl$$

Temperatures of 1000 to 1200°C were employed to produce an alpha alumina deposit at reasonable rates on substrates such as 5-mil tungsten. Other substrates included platinum, tantalum, silicon carbide, graphite, and boron. Despite many variations in experimental conditions, most of the filament coatings of alpha alumina were poor. Interestingly enough, a corncob structure similar in appearance to that found for boron filaments was produced and found to be theta-alumina, which was converted to alpha alumina after a few minutes of a 1930°C heat treatment in hydrogen. The fibers made were very weak. Passing a tungsten wire through molten aluminum isopropoxide $(C_3H_7O)_3$ Al, which has a melting point of 118.5°C, produced an alpha alumina coating at 1000°C.[31]

This coating contained some dissolved aluminum isopropoxide in the reaction deposit. Some chemical attack, and subsequent breakage of the tungsten substrate, was observed. Alpha alumina was also coated on silicon carbide substrates through a molten isopropoxide bath.

Another recent effort has been the investigation of drawing a glassy fiber which is then converted to a high alumina filament. This makes fast-drawing feasible and thus attractive as an economical method to produce continuous oxide filaments. Simpson[32] utilized compositions based on the $Al_2O_3 \cdot B_2O_3 \cdot P_2O_5$ ternary system to prepare glassy filaments from melts containing high concentrations of alumina, up to 60%. Tensioning the glassy filament eliminated macrowarpage and microkinking associated with the thermal conversion of the aluminum-boron phosphate glass filaments to polycrystalline high alumina filaments. The best glass composition for drawing was found to be $50Al_2O_3 \cdot 25B_2O_3 \cdot 25P_2O_5$. Minor additions of yttria and thoria increased the working range of the glass but tended to inhibit devitrification. The filaments were converted to polycrystalline alumina from the glassy phase by utilizing rapid heating in the relatively low vacuum achieved by ordinary mechanical pumps, for example, 2×10^{-2} mm Hg to 1.5×10^{-1} mm Hg. Heating cycles from ambient to 3000°F (1649°C) at heating rates about 50°F (10°C) per minute were used. The results of heat treatment on tensile strength of converted filaments are shown in Figure 19.[32] Holding the filament at a given temperature for long periods of time (5 to 60 min) did not significantly increase strength levels. In some cases

FIGURE 19.   Tensile strength as a function of filament conversion temperature when heated in 0.13 mm H_2 at 50°F/min.

traces of a second phase were present as small particles in the grain boundaries, indicating incomplete loss of phosphorus and boron. Strengths achieved were disappointing, with some increase noted when laser beam heating was used to convert the glassy phase to alumina. The highest strength at 65,000 psi was disappointing compared to currently available melt-grown filaments.

The most encouraging recent advance in reinforcements for metal matrix composites is undoubtedly the commercial development of a wide diameter boron filament. This filament and the silicon carbide coated analog in wide diameter Borsic have much improved transverse properties compared to the thinner predecessor boron and Borsic filament. This results in major improvements in the properties of metal composite structures such as B/Al and B/Ti in the transverse direction to the axis of reinforcement. It considerably reduces the cost of the filament; for example, late 1971 estimates for 4-mil boron filaments were $210 per pound, while estimates for large-scale production of 5.6-mil boron were given at about $100 per pound. This is primarily a result of reducing the percentage (and cost) of tungsten substrate since the same tungsten wire is used in both size filaments. The wider diameter filament is easier to handle, reportedly breaks and splits far less, and has a better surface consistency (fewer obvious defects which result in premature breaks) than the thinner filaments. B/epoxy and B/Al tapes are also more economical with the wide diameter reinforcement. The properties of these filaments currently commerically available[33] are given in Table 2. Unfortunately, many design data and performance specifications are available for 4-mil boron which should be reevaluated for given metal matrix composites with the improved 5.6-mil filament. This will take considerable industrial effort and should be encouraged whenever possible. The 5.6-mil boron is easily obtained by utilizing the same production reactors on which 4-mil boron is made with minor modifications and changes in deposition conditions.

TABLE 2

Boron Type Filament Properties[33]

Property	Unit	Borsic		Boron	
		5.7 mil	4.2 mil	5.6 mil	4.0 mil
Diameter	10^{-3} in.	5.7 ± 0.1	4.0 ± 0.2	5.6 ± 0.1	3.9 ± 0.2
Ultimate tensile strength	10^3 psi	$450 - 475$	$425 - 450$	$450 - 500$	$450 - 500$
Minimum tensile strength	10^3 psi	300	300	350	335
Modulus of elasticity	10^6 psi	$59 - 60$	$57 - 60$	$59 - 60$	$57 - 60$
Density	lb/in.^{3}	0.092	0.096	0.090	0.095
Length per pound	ft	34,500	57,000	38,000	70,000

Chapter 3

CONSOLIDATION PROCESSES

A wide variety of metallurgical processes have been employed for the fabrication of filament reinforced metal matrix composites. The techniques differ in great detail from those which employ conventional powder metallurgy and slip casting methods to newer metallurgical techniques such as diffusion bonding and plasma spray bonding, and finally to specially adapted techniques of hot roll bonding and liquid-metal infiltration. The greatest attention has been placed on diffusion bonding and, more recently, spray bonding as the preferred methods. Eventually, methods adaptable to continuous processes or casting methods would be expected to predominate as emphasis shifts from research to establish composite system properties to development of economical structures for the commercial market. Indeed, substantial progress must be made in establishing economical fabrication techniques before composite structures will become competitive with other materials they seek to replace. Hence, this is a most important topic of concern in the metal matrix technology assessment.

Each process employed to consolidate a composite structure requires specific methods to deal with the highly anisotropic materials and the relatively fragile reinforcing filaments which differ from conventional metallurgical techniques. Each process employed must meet to some reasonable degree the following objectives:

1. Incorporate the filament without breakage.

2. Consolidate the composite with minimal filament degradation.

3. Establish and maintain filament alignment.

4. Offer variable filament volume loading.

5. Achieve a high density matrix.

6. Establish filament-matrix interfacial bond sufficient to transmit applied load from matrix to filament.

7. Allow a degree of matrix selection and alloying.

8. Allow for post-fabrication heat treatment.

9. Give flexibility in filament spacing.

10. Some methods must provide capability for cross and angle ply lay-ups.

11. Minimize product variability.

12. Be amenable to scaling up.

Those methods which are considered for production items must also be reasonably economical and rapid and have flexibility of form of material to be handled and shapes to be produced. No one process satisfies all counts and often several objectives cannot be met at once. For example, incorporating a filament with a foil may require high temperatures so that metal will flow around the filament without breakage, and high pressures to achieve a high density composite without large voids. This set of conditions often results in severe degradation of the filament and even the formation of deleterious interface reaction products.

The compatibility problem in metal matrix composites is of sufficient importance that factors related to controlling filament-matrix reactions and attendant interface phenomena are treated in a separate section. The optimization of fabrication parameters, particularly on the extensively studied B/Al system, has been treated within the compatibility context. In this section attention is placed mainly on the physical nature of the various processes employed, some typical examples of where these processes have been used, and the results obtained. Some advantages and disadvantages have been reviewed, but in many cases the processes have not been optimized and one should be slow to draw conclusions on the efficacy of some of them. In much of the work insufficient information is available on the variation of mechanical properties of the composites with processing parameters. Only a general idea of feasibility has often been demonstrated. As the field of composite technology develops, however, the emphasis on process techniques has been narrowing down to methods such as diffusion-bonding, where a great number of encouraging results have been achieved and documented.

The consolidation processes used for metal

matrix composites have also been reviewed by Alexander, Shaver, and Withers,[2] Davis and Bradstreet,[34] and by Snide, Lynch, and Whipple.[8]

Diffusion-bonding

The most widely used method of consolidation of metal matrix composites is the simultaneous application of heat and pressure, known as diffusion-bonding. This process may be used to consolidate tape or wire preforms made by other methods such as plasma spraying or hot roll bonding, or to consolidate filament-matrix foil arrays. It is a static pressure process as differentiated from high energy methods which use dynamic pressure techniques. This method is used commonly to consolidate B/Al, Borsic/Al, Ti/SiC, and Ti/Borsic composites. This dates back to the early work on B and Borsic/Al[35,36] through more recent work[37-39] and on Ti/SiC.[40]

The diffusion-bonding process for a single uniaxial filament-reinforced metal matrix is shown schematically in Figure 20.[8] The process has been used in some form to fabricate most of the commercially available metal matrix composite materials. The lay-up shown in Figure 20 is for a simple uniaxially aligned filament reinforced metal foil composite. The diffusion-bonding step requires specific conditions for each type

composite. For example, the maximum conditions required for typical diffusion bonded composites are 850°F (454°C) for 1 hr at 6,000 psi for B/Al,[40] 1600°F (871°C) for 1 hr at 6,000 psi for SiC/Ti,[40] and 1022°F (550°C) for 1 hr at 5,500 psi for plasma-sprayed tape of Borsic/Al.[39]

Some typical variations on the "preform" materials used in diffusion-bonding are indicated in Figures 21 and 22. Simple foil-filament arrays shown in Figure 21 are formed by careful winding of accurately spaced filament onto a foil-covered drum and holding the filament together with a resinous binder which is fired out in subsequent processing steps.[41-43]

Control of filament spacing can be achieved by co-winding of the matrix wire between the reinforcing filaments. This also reduces the amount of excess bulk which must be removed in the consolidation process. Although foil-filament arrays are commonly employed, the removal of 35% or more porosity, the decomposition of the resinous binder, and the maintenance of consistent filament spacing are considerable problems. The incorporation of matrix wires by co-winding (more properly, simultaneous winding) is more complicated initially but substantially improves subsequent consolidation efforts.

The plasma-spray bonding process will be

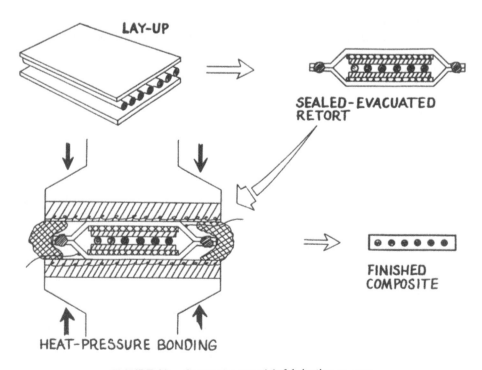

FIGURE 20.   Composite materials fabrication process.

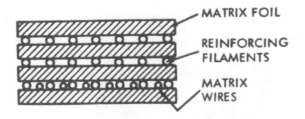

FIGURE 21A.　Metal foil-filament array.

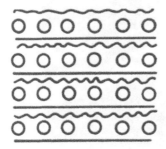

FIGURE 21B.　Stacked metal matrix tape preform.

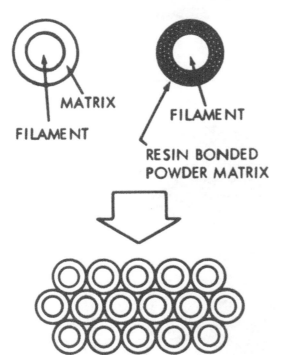

MATRIX

FILAMENT

FILAMENT

RESIN BONDED
POWDER MATRIX

FIGURE 22.　Matrix coated filament array.

considered separately in more detail. As shown in Figure 21, tape preforms provide excellent starting materials for the diffusion-bonding process and are often utilized.[44-46] The tapes are easily handled,

filament spacing is maintained, and there is no binder problem. Internal porosity and surface roughness of plasma-sprayed tapes present problems in consolidation in the production of low-porosity composites. Complex cross-ply and angle-ply orientations are conveniently obtained by alternating the preform tape layers. The cost of producing monolayer tapes is one distinct disadvantage to the process.

Figure 22 illustrates the use of matrix-coated filaments to form uniform uniaxially oriented filament matrix arrays.[47,48] Matrix coating can be achieved by liquid or vapor deposition techniques, by electrolytic deposition of a thin metal layer, or by powder metallurgy with a resin-bonded metal powder. There is a porosity problem here as well, and the process tends to be experimentally difficult since the coating problem itself on a larger scale basis is a major concern. The chief advantage would appear to be in very high volume loadings where filament deformation and overlap becomes a serious problem.[2]

Two mechanized techniques for preparing aligned filament mats are shown in Figures 23 and 24.[49] An apparatus for filament winding is shown in Figure 23. The filaments are wound on a mandrel mounted on a lathe. A loom, similar to that used for weaving cloth, can be employed to prepare filament-metal ribbon mats as shown in Figure 24. These processes are used to control filament spacing for the preform materials. The weaving process is more complex because it introduces additional interfaces in the fabricated composites and mechanical damage to the filaments during weaving is a major problem. The mandrel winding is usually employed for preparing aligned filament mats when tape preforms or grooved foils are not used.

Using diffusion-bonding as the most advanced consolidation process, marked improvements in the properties of B/Al composites have been achieved. Optimizing the fabrication parameters, utilizing the wide diameter 5.6 mil boron filament, and heat treating the composite material have produced excellent results. Maximum values for 48 v/o 5.6 mil B/6061 Al have been reported (uniaxially reinforced, heat treated) as 220,000 psi longitudinal tensile strength, over 40,000 psi transverse tensile strength, and 625,000 psi compressive strength.[50] More conservative figures are used in determining design allowables for current large component studies such as on

FIGURE 23.   Mandrel winding setup.

FIGURE 24.   Automated hand loom for weaving filament mats.

fuselage bulkheads.[51] However, the progress is very impressive and the consolidation process by diffusion-bonding very well established. The size of individual panels achieved has reached 10 ft x 6 in. x 1/4 in. for B/Al.[50]

Plasma Spray Bonding

The method of plasma spray bonding to prepare monolayer tapes which are then consolidated by diffusion-bonding into thicker sections has become an important consolidation process.[35,36,38,39] The use of silicon carbide as a protective coating on boron (Borsic filament) yields sufficient protection during the plasma spray procedure for the preparation of Borsic/Al composite tapes.

The plasma spray apparatus is illustrated in Figure 25. A cylindrical mandrel, which may be of considerable size, containing a layer of matrix foil or plasma-sprayed matrix material is overwrapped with an aligned layer of reinforcing filaments with controlled spacing of the filaments through filament winding techniques. The mandrel rotates in front of the plasma arc which thus deposits an even layer of matrix material in the form of tiny liquid droplets which solidify rapidly on impinging the mandrel material. The process is conducted in argon or with the substrate shrouded in the effluent arc gas, partially shielding the substrate from oxygen. The molten zone contact time with the filaments is sufficiently short to eliminate serious filament degradation. There is considerable oxygen pick up, however, and the as-sprayed matrix is not dense.[2] Typical arc conditions are 30 V, 400 to 500 A, and 140 to 160 ft^3/hr argon flow.

The spray deposit is affected by the rate of powder feed which affects both deposit efficiency and the quality of matrix-filament bond achieved. Good results were achieved at feed rates of 3 lb/hr of metal powder, −240 to +400 mesh, with a 4- to 5-in. torch to substrate distance.

Advantages of the process include the "locking in" of good fiber spacing if the initial monofilament winding technique is correct and the excellent flexibility of monolayer tapes in hardware fabrication. Complex parts and angle-ply configuration are possible without prohibitive monofilament handling problems. Large panels are no particular problem either. Disadvantages include the cost of intermediate processing to a monolayer tape, although the process of plasma spraying is not too expensive per se, and is relatively simple. Limitations of the method to relatively low melting materials and oxygen contamination which eliminates practical consideration of metals like titanium are severe, and the principal interest thus far is only on aluminum matrix composites.

Monolayer tapes can be consolidated at lower pressures and temperatures by utilizing braze bonding. The use of Borsic/Al tapes consolidated with aluminum brazing, alloy foil was investigated[37] and shown to offer advantages of lower pressure and lower temperature bonding to other materials in joining operations. The braze alloy was successfully incorporated into the composites to give properties approximating conventionally diffusion bonded composites.

Plasma spraying has been conducted by Greening[52] to incorporate tungsten fiber into a tungsten matrix. Feasibility for the method was indicated in the results for economic fabrication of high temperature refractory composite structures in either simple or complex configurations on conventional equipment.

Electroforming

Electrodeposition as an approach to fabrication of metal matrix composites was one of the first methods studied because the composite could be formed at low temperatures, thus minimizing the degradation of reactive filaments in metal matrices. The electroforming method has been described in a number of investigations.[53-57] A schematic of the electroforming process is shown in Figure 26.[8] The metal is deposited on a mandrel from a plating bath while the filaments are wound concurrently on the mandrel. Boron filament-

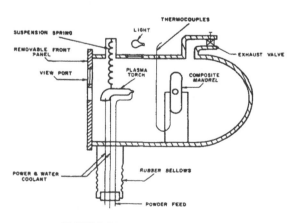

FIGURE 25.   Plasma spray chamber.

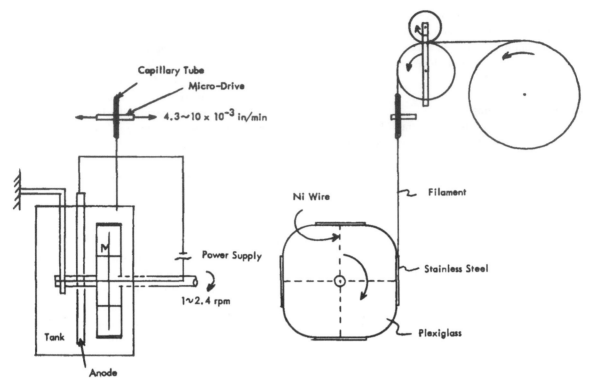

FIGURE 26. A schematic diagram of the winding apparatus.

reinforced nickel matrix composites have been fabricated successfully using this method and satisfactory properties have been obtained.[2,8,53] The boron-nickel system, however, is of little practical interest because of the rapid interaction at high temperatures and the low melting eutectic temperature. Two practical limitations which seem to be inherent in the electroforming process are (1) the impurities which are incorporated into the composite from the bath during deposition and (2) the inability to deposit alloys from solution.[8] The schematic shows a cross section and end view of the mandrel, illustrating the winding from a filament guide of evenly spaced boron filament and the solution deposition bath which makes the mandrel the cathode for the nickel plating from the solution.[2]

A number of advantages can be cited for this process, including:

1. A low room temperature fabrication process.

2. A high density monofilament matrix can be obtained.

3. Excellent filament-matrix interface contact is possible.

4. Filament spacing is accurately controlled.

5. Volume loading of reinforcement is quite flexible.

6. A variety of mandrel shapes can be accommodated.

In actual use additional practical limitations appear[2] which make the deposition of multiple layers extremely difficult. For preparation of monolayer tapes containing up to at least 45 v/o filament, little void problem is encountered. Thus, as a source of monolayer tapes for further consolidation by diffusion bonding, the method has reasonable attractiveness. Figure 27 represents the deposition of a monolayer and then multilayer composites as observed with experimental electroforming on the B/Ni system.[2] Voids form where the surface contour becomes such that the growth from the filaments (deposition of nickel) meets before the growth from the substrate nickel on the mandrel has proceeded sufficiently around the filaments. The process is made more critical with fairly high volume loadings of filaments in multilayers. Unless the surface contour from one layer to the next is exactly conforming to the succeeding layer, voids appear underneath the

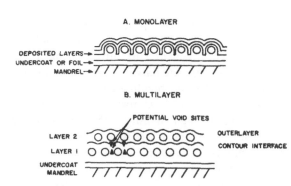

A. MONOLAYER

DEPOSITED LAYERS →
UNDERCOAT OR FOIL →
MANDREL →

B. MULTILAYER

POTENTIAL VOID SITES

LAYER 2
LAYER 1
UNDERCOAT
MANDREL

OUTERLAYER
CONTOUR INTERFACE

FIGURE 27. Surface contour of electroformed composite.

individual filaments, and between the filaments in the same plane. Variations in filament spacing change the contour and require exact adjustments in the deposition rate. This means that flexibility in spacing and volume loading is really not achieved without considerable effort. Any misalignment or spacing variation may lead to increased void formation.

The electroforming method has been combined with sintering techniques by Donovan and Watson-Adams[58] to prepare carbon fiber reinforced nickel composites. The method they utilized was to electrodeposit nickel on aligned carbon fiber bundles and then subsequently to sinter the coated bundles into a composite structure. It was found in this study that producing carbon fiber reinforced electroformed composites without benefit of subsequent compaction and heat treatment was not practically possible. This substantiates previous efforts on the difficulties of depositing multilayers by this process. The process of plating and sintering under pressure was judged by Donovan and Watson-Adams to be useful to produce composites up to 45 to 50 vol % carbon fibers. Upon heat treatment of the composite at 1150°C (2102°F), grain growth of the carbon fiber resulted which would deteriorate the mechanical properties of the composite. Actually, this work refers to other work that indicates that 1000°C (1832°F) or less may be the limit for this reaction. It was found necessary to heat treat the composites at temperatures approaching this limit to eliminate the weak bonding planes where electroformed layers connect between fiber bundles. Model systems of steel wires in nickel and carbon rods in nickel were also studied to establish critical plating conditions and solution conditions for the system. In

conclusion, the method has marginal utility and serious problems for most metal matrix composites.

Liquid Metal Infiltration

Where possible, liquid metal infiltration has been used to consolidate metal matrix composites. Illustrative of this approach is the work of Jackson et al.[48] on the melt coating of SiO_2 fibers by passing single filaments through beads of molten aluminum prior to hot pressing bonding to form SiO_2/Al composites. Several model refractory metal systems have been consolidated in this manner such as with tungsten reinforced copper, where little or no mutual solubility exists. The tungsten wires collimated in a ceramic tube were infiltrated by liquid copper.[59] The technique is limited and has not received wide use because of the few reinforcements which are stable in molten metals. This technique has been employed on Al_2O_3 whisker reinforced composites. The problem with the Al_2O_3 whisker in matrices such as Al and Ni has not been whisker degradation reactions, but rather the proper wetting of the whisker with the matrix. The wetting is controlled by various coatings, matrix alloying, and control of the infiltration atmosphere.[8]

Figure 28 shows a typical glass system[8] used in the laboratory to infiltrate liquid metal into pre-aligned reinforcements. The vacuum casting method is shown using an induction heated graphite crucible to melt the matrix material, which then flows over the filaments in a ceramic mold. This type of apparatus is limited to a batch process and relatively small size specimen materials. It can be modified, however, to provide a continuous casting process where the length of the cylindrical or other shape cast rod can be made at any reasonable length up to that of the reinforcing filament. This is where the major interest in liquid infiltration has been centered.

Using a tubular arrangement similar to Figure 28, the filaments are collimated into the crucible with the powder melt and continuously drawn through an exit orifice at the bottom of the crucible and then through cooling coils and a long freeze tube.[2] In practice it is simply passing the bundle of filaments through a metal bath so that the filaments are wet with metal, and removing the excess metal as the bundle is drawn through the orifice. Schuerch[60] early showed that it was feasible to use liquid metal infiltration to cast a

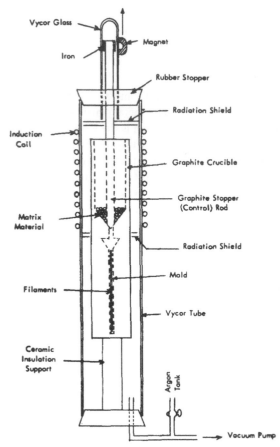

FIGURE 28. Schematic diagram of the apparatus for vacuum casting.

key flexibility to the process. The length of continuous rods by this process is limited only by the length of filament currently available (over 10,000 ft). Composite bars were produced by this method by diffusion bonding an array of rods of B/Mg or Borsic/Al.[63] Almost perfect hexagonal packing in a high density, 75 vol % B/Mg composite rod has been reported.[2] The continuous casting process is capable of yielding any uniform cross section of a uniaxially aligned filament reinforced composite. The reinforced magnesium alloys made by this process are considerably higher in specific strength and specific modulus than existing aluminum and titanium alloys. There are the usual disadvantages to magnesium, such as corrosion susceptibility, however, and having only uniaxial reinforcement since angle-ply or cross-ply configurations could not be made easily in this manner.

High Energy Rate Forming

Very high pressure pulses of short duration have been utilized to fabricate composites in a method known as high energy rate forming. Much of this work has been pioneered by Williford and Snajdr.[64] One device for this type of fabrication which was employed to consolidate whisker-reinforced metal matrix composites is shown in Figure 29.[8] The method draws its advantage from the instantaneous application of very high

boron-reinforced magnesium where there was sufficient stability of boron in contact with the molten metal. Later, Alexander and Davies[61] demonstrated the continuous casting process and obtained excellent results in fabricating boron-reinforced magnesium rods in this manner.

On the other hand, Wolff[47] was unable to fabricate filament tapes satisfactorily by passing boron filament rapidly through molten aluminum. Serious degradation of the filament occurred with only a brief contact time with the aluminum melt. Still more recently Davies, Shaver, and Withers[62] have reported the development of the continuous casting process for Borsic/Al as well as for B/Mg. Magnesium alloys prepared by them have been formed in several shapes including I-beams, circular cylindrical rods (both hollow and solid), and angle beams. The filaments can be placed uniformly within the matrix and selectively in particular sections such as the flanges of an I-beam, according to this effort, thus imparting

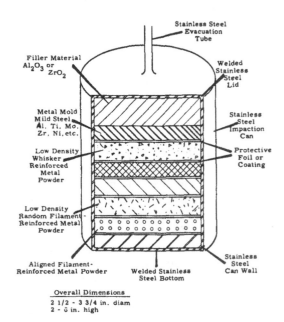

FIGURE 29. Composite billet for high energy-rate forming composite plates.

pressures for very short time periods which prevents interaction of matrix and filament or whisker reinforcement while providing densification of reasonably complex shapes. In actual practice, however, filament breakage at low temperatures forces the use of higher consolidation temperatures where serious filament degradation occurs.[2] For whiskers the control of alignment and distribution is insufficient and secondary processing is needed.[8] For the SiC/Ti system a temperature of 1100°C for the high energy rate forming process is much above the 900°C required for diffusion-bonding and creates reactivity problems.[2]

Typical conditions for this process are applied pressures of 400,000 psi and an energy pulse duration of micro- to milliseconds. The apparatus may utilize explosive, electromechanical, or pneumatic-mechanical means to achieve the high pressure pulse. The work by Williford and Snajdr has employed a Dynapak machine which is a pneumatically assisted mechanical impaction device.[65] A set pressure piston activates the punch and ram assembly by pressurized gas and accelerates the punch into a die and against the fixed work piece with a velocity of 66 ft/sec at impact. The duration of the initial pulse is 3 to 15 ms in this device, depending on the material on which the forming operation is conducted. The billet assembly which contains the composite is heated prior to the compaction step. The so-called "HERF" method was used extensively as a production tool for cermet and nuclear materials densification and therefore has considerable background as a possible production technique.[65] The potential of the method for scaling up to large size plate or sheet materials with reproducible properties is considerable, providing essential research on the appropriate processing conditions and on problems such as filament alignment and distribution is conducted. At present little additional work is being done.

Earlier work on whisker-reinforced composites was conducted on systems such as SiC/Ti-6Al-4V, Al_2O_3/Ti-6Al-4V, Al_2O_3/Ni, and Al_2O_3/Fe-25Cr-5Al with very unsatisfactory results.[65] These systems are simply not ductile enough to employ the high energy pulse densification without excessive breakage. Bonding is also poor, as it has been with sintered whisker-metal matrix composites. In this case chemical compatibility and matrix microstructural control as demon-

strated by Robinson[66] are insufficient. Preliminary results with continuous filament composites using SiC/Ti-6Al-4V produced reasonable mechanical properties, especially considering the high temperatures employed. A temperature of 1000°C was the minimum necessary to achieve densification[65] and the interaction layer thickness was found to be 6500 Å, which is greater than desired to minimize loss of transverse strength. This would account for the rather low transverse strengths that were reported and is quantitatively illustrative of the difficulty in utilizing the method.

Explosive welding has also been investigated as a method for composite fabrication.[67] In the explosive welding process the chemical explosive is used to impinge mating surfaces at high velocities, developing high pressure at the collision interface. The explosive bonding parameters of explosive weight/unit area of cladding surface and initial interface clearance are controlled so that the dynamic angle between the colliding surfaces is maintained within certain critical limits. A portion of the colliding surfaces which is called a jet becomes fluid or gaseous, and is expelled. The collision of the surfaces and their subsequent deformation, in addition to the jet action, removes or breaks up the surface film, giving film-free surfaces in intimate contact. Using a non-nitroglycerin dynamite containing nitrostarch and ammonium nitrate, a dry powder with a detonation velocity of 8,000 to 12,000 ft/sec has been employed. The explosive cannot be detonated easily and is relatively inexpensive. Foil filament layers have been explosively compacted with AFC 77 steel wire-reinforced aluminum alloys. It was necessary to heat treat the formed composite to develop a suitable metallurgical bond.[67]

A number of other systems have been reported as successfully fabricated, such as beryllium-reinforced aluminum.[68] A uniform microstructure and negligible fiber fragmentation were observed utilizing an ammonium nitrate-nitrostarch explosive.

Successful compactions of other systems include fiber-powder compacts and copper-coated steel wires consolidated by a sheet explosive containing PETN.[67] The results of explosive welding are on a sufficient number of systems to encourage further research on this processing approach.

Hot-Roll Bonding

The simultaneous application of heat and pressure in the diffusion-bonding process has been modified to provide for a short reaction time during fabrication in a process known as hot-roll bonding. This method has been extensively employed by Metcalfe and his co-workers [46,69-71] to counter the reactivity problem with reactive systems such as B/Ti. Valuable insight into the influence of filament matrix reactions on the mechanical properties has been obtained utilizing this method and is discussed under the treatment of compatibility problems. The roll bonding method provides for a very short time of contact between the filaments and matrix at temperature, but is practically limited to tapes as monolayers or a few layers thick. These tapes are then diffusion bonded into larger sections at lower temperatures and pressures than the initial bonding conditions since only a simple surface metal-to-metal bonding is involved. The apparatus for hot-roll bonding is indicated schematically in Figure 30. In a typical diffusion-bonding procedure[46] the boron filaments were fed from supply rolls into precisely set positions to align in the grooves of a titanium foil as center tape and two grooved cover tapes fed above and below the center tape to form a sandwich structure. This was fed between two molybdenum rollers which were resistance heated to $1800°F$ ($982°C$). The contact time was in the order of 1 sec and the resulting two filament layer tape had a reaction zone width of about 500 Å, which is excellent. In a typical fabrication run 30 filaments were aligned in a 0.1 in. wide portion of a 0.5 in. wide tape which was 13 mils thick. This gave a 25 vol % loading.

A major problem in hot-roll bonding is the difficulty in obtaining high volume loading of the filaments. Another is cross- and angle-ply configurations which, for diffusion bonded sheets in alternate directions, are reasonably straightforward. For thin tapes this is a big problem in consolidation. To make wider and thicker tapes requires a degree of temperature and pressure control on the rollers that would inhibit large scale up procedures. The grooving required for alignment is also a substantial consideration. Nevertheless, the success with highly reactive systems makes roll bonding attractive as a potential fabrication technique.

Other Methods

Matrix-coated filaments as shown in Figure 22 for a preform in the diffusion-bonding process have recently been employed in continuous drawing and rolling experiments.[72] Composites of 50 vol % B/6061 Al were sheathed in aluminum and a series of cold-drawing, and cold-rolling, hot-rolling, hot-drawing, and diffusion-bonding experiments was conducted. A variety of volume loadings were achieved by changing the cladding thickness of aluminum from 0.5 to 30 mils. Although much needs to be done, the results indicate that with adequate sheathing, cold-drawing can be achieved without filament fragmentation. In continuous processing the interfilament spacing control is a substantial problem. Advantages of matrix-coated filaments in handling are considerable. They can be bent, consolidated into sheathed bundles, manipulated without filament surface damage, and the bundles cold-pressed into compacted preforms of varying shape without filament fracture. For the rolling experiments 85 matrix-coated filaments were placed in a 6061-T6 Al sheath of 0.125 in. outer diameter and 0.030 in. inner diameter. The cylindrical preforms were cold-rolled through 1.75 in. diameter rolls in 21 passes to achieve a 0.042 in. thick flat preform for a 66% reduction. The flat preform was subsequently hot-rolled at $1000°F$ ($538°C$) and $1060°F$ ($571°C$). In one pass at each temperature further reductions of 9.4% and 24%, respectively, were achieved.[72] Cold-rolling to this degree produced readily detectable voids showing a need for process improvement. Hot-rolling improved the densification process and showed the need for edge restraints to maintain filament spacing. In view of the exploratory nature of this work the results are quite promising. For cold-drawing experiments a 6061 Al-T6Al sheath of 0.093 in. outer diameter and 0.066 in. inner diameter containing 38 matrix-coated filaments was swaged down at one end and drawn through a 0.088 in.

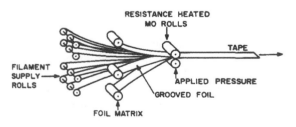

FIGURE 30.　Hot-roll bonding.

diameter orifice. A 10% reduction in area at room temperature was achieved with filament alignment maintained. Several subsequent passes through successively smaller orifices to achieve additional reductions up to 10% each were made.[72] Some void formation was observed in cold-rolling but interfilament spacing was maintained. Hot-rolling experiments were conducted at 970°F (521°C) and 1000°F (538°C) but were not successful due to an accumulation of aluminum at the orifice. Diffusion bonding of matrix-coated filaments at 5000 psi for 90 min at 975°F (524°C) indicated no particular advantage over other methods of consolidation.

Chemical vapor deposition has been employed to infiltrate filament arrays to form metal matrix composites. Withers,[73] for example, has formed Be/Al composites by infiltrating beryllium filaments wound on a heated mandrel with aluminum deposited by thermal decomposition from aluminum alkyls. Subsequent layers after the first chemically deposited layer tend to be irregular and voids appear at the contact surfaces between layers. In this manner the electroforming and chemical vapor deposition processes have the same serious problem. The method has the advantage of being conducted at relatively low temperatures. Thus the consolidation of W/W or B/W composites is interesting as demonstrated by Greszczuk.[74] Temperatures of 800°F (427°C) to 1100°F (593°C) were employed which resulted in consolidation without the reactivity problems attendant with the higher temperature processes such as diffusion-bonding. This is inherent in the process since the metal-organic compounds with the volatility to be considered for chemical vapor deposition have relatively low decomposition temperatures compared to the bonding temperatures for most of the metals of interest. An inherent disadvantage of adequate control of a complex chemical vapor deposition process and the cost of the method may preclude more serious attention than the feasibility studies thus far conducted. Alexander, Shaver, and Withers[2] feel that continuous tape formation by chemical vapor deposition is feasible as a consolidation process.

Cold pressing and sintering and powder slip casting have been used mainly for metal filament-reinforced metal composites to form uniaxially reinforced structures. Where extreme stability of composite components is available the method has the advantage of widely available equipment and controlled processing. Because of the relatively high temperatures and long times at temperature and pressure, the method is not attractive for most systems of interest and has not been widely employed.[2]

Extrusion and rolling techniques were employed by Baskey[75] to consolidate refractory metal wires in nickel base superalloys. Randomly oriented discontinuous filament-powder matrix mixtures were cold or hot pressed into a preform for extrusion to yield uniform cross section rod composites containing aligned fiber reinforcements. Effects of preform treatments and extrusion parameters and post extrusion heat treatments on mechanical properties were examined. Results were compared with the unreinforced matrix and to wrought forms of the alloys which are the competitive materials to these composites.[2] Improvements were obtained in both nickel base superalloy and in titanium alloy composites in tensile and yield strengths and stress rupture properties formed by extrusion of discontinuous refractory metal filaments and powder metal matrices. As in later studies, continuous filament composites of equivalent volume loading reinforcement had better properties than discontinuous filament composites.[2,75]

Hot extrusion of B/Al was examined by Meyerer.[76] He also reviewed some of the previous efforts at extrusion as a consolidation process. The process employed in his work is shown schematically in Figure 31. The filaments in the composite bar to be extruded were aligned by hand in the Al powder layer and then cold pressed into a preform. This was placed into a 2024 Al can and the can filled with other preforms and tightly fit with 2024 Al shims. The sealed can was preheated at 800°F (427°C) and the hot billet extruded with a 10:1 reduction ratio at an extrusion rate of 1.6 in./min. "Continuous" filaments of 3-in. length and "discontinuous" filaments of 1-in. length were loaded to relatively small volumes up to 12% continuous filament reinforcement and 6% discontinuous filament. The extruded bar (48 in. long from a 5.45 in. can on a typical run) was sectioned for metallographic examination and tensile testing. Some problems in void formation, fiber misalignment, and fiber damage were noted, but could undoubtedly be controlled with a sophisticated alignment and extrusion technique. With the low volume reinforcement employed the mechanical property changes were rather small. A significant

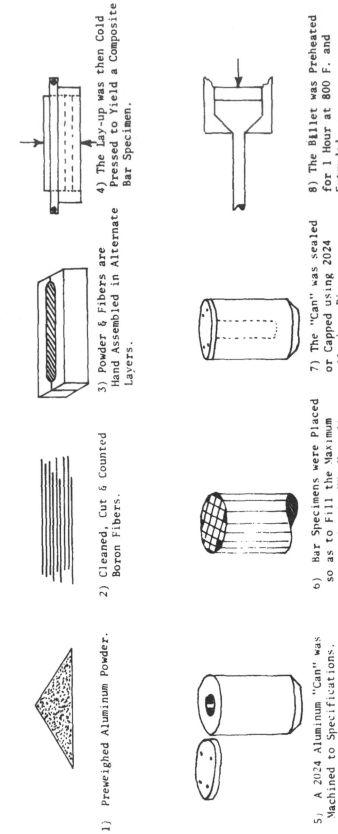

1) Preweighed Aluminum Powder.

2) Cleaned, Cut & Counted Boron Fibers.

3) Powder & Fibers are Hand Assembled in Alternate Layers.

4) The Lay-up was then Cold Pressed to Yield a Composite Bar Specimen.

5) A 2024 Aluminum "Can" was Machined to Specifications.

6) Bar Specimens were Placed so as to Fill the Maximum Volume in the "Can" cavity. Remaining Space was Fitted with Shims of 2024 Aluminum.

7) The "Can" was sealed or Capped using 2024 Aluminum Pins.

8) The Billet was Preheated for 1 Hour at 800 F. and Extruded.

9) The Resultant Bar was Sectioned into 1" and 3" Specimens.
   a) 1" Specimens for Metallography
   b) 3" Specimens for Mechanical Testing

10) The 3" Specimens were Machined into tensile Specimens of the Above Configuration.

FIGURE 31.   Schematic diagram of extrusion process.

modulus increase was obtained, however, which was consistent with a rule-of-mixtures prediction. Strength increases were nominal and offset by elongation decreases. The study does indicate feasibility but much additional work would be required to show practical utility for this type of extrusion process for brittle reinforcements such as boron. High volume loadings of reinforcement would probably cause extensive filament damage. The matrix-coated filament would appear to offer more potential in this regard for the extrusion process.

Chapter 4

COMPATIBILITY AND FABRICATION

Filament-matrix compatibility includes the familiar chemical interactions between matrix and reinforcement during consolidation, secondary processing steps, and in use. Generally, filament-matrix consolidation steps require more severe conditions than anticipated use conditions, but there are exceptions in which low temperature *fabrication techniques, such as electroforming and* explosive forming, are employed. Compatibility is a serious problem in metal matrix composite development.[4,8,11] In a broad sense it also involves mechanical compatibility, such as thermal expansion mismatch effects and the synergistic effects which are possible[77,78] between components. Stuhrke[79] has demonstrated these in diffusion-bonded boron-reinforced aluminum where he suggested that Poisson's ratio differences between the components result in significant *transverse stresses which put the matrix in bal-* anced biaxial and, possibly, triaxial tension. When an adequate filament-matrix bond was attained, he demonstrated a 20 to 30% increase in axial tensile strength and elastic modulus over the theoretical rule-of-mixtures values. Compatibility, therefore, can have several connotations, all correct in their own way. While gross chemical reactions are a problem, some degree of reaction appears necessary to promote bonding. What, then, constitutes the bonding characteristics that minimize the degrading influence of reactions and yet allow efficient load transfer between composite components?[5] Answering this question requires more *definitive characterization of specific metal-* reinforcement interfaces than has been obtained thus far for most metal composites.

In studying this basic problem of prevention or control of chemical interactions which will degrade either the filament or the matrix, the following approaches have been pursued:[5,8]

(1) develop *new reinforcements which are* thermodynamically stable with respect to the matrix,

(2) develop diffusion-barrier coatings which reduce the filament-matrix interaction,

(3) develop alloy additions which reduce the

activity of the diffusing species and enhance the physical compatibility, such as minimizing the thermal expansion mismatch at the interface.

In connection with these criteria it must be remembered that thermodynamic considerations offer a suitable starting point for selecting stable systems, but the kinetic data on reactions and diffusion rates will define the practical thermal stability of a given composite system.[4]

Gross Reactions

Much effort has been expended to find systems which do not undergo gross reactions at consolidation and potential use temperatures. A great *number of systems were found to be quite* reactive.[4,8,11,22,80,81] For example, investigations of a SiC-reinforced Ni-Cr alloy for oxidation-resistant composites at elevated temperatures showed this system to be gross-reaction limited. Phase compatibility data are now available which establish definitive limits for this system and the identity of various reactions that occur.[82,83] The data were unavailable when this metal matrix composite effort was initiated. Often crucial phase stability and thermodynamic data are not readily available in useful form when the systems are investigated. Systematic screening of potential systems is commonly employed to establish the degree of reactivity and the potential for alloying or coating protection to reduce interactions.

An oblique view of the filament-matrix interface in a SiC-reinforced titanium composite is shown in Figure 32.[84] The multiplicity of distinct zones is indicative of the complexity of interactions that occurred. At temperatures of 650°C compounds definitely formed included Ti_5Si_3 and TiC, but other silicides of titanium are quite likely since titanium forms several silicides.[4,29,81]

In Figure 33 the reactions between several filaments and iron are shown after a 1 hr anneal (100 hours for TiB_2) at 900°C.[81,85] In each case the filaments were deposited on a tungsten core substrate. The reactions show complex interaction-diffusion patterns, particularly for the B/Fe

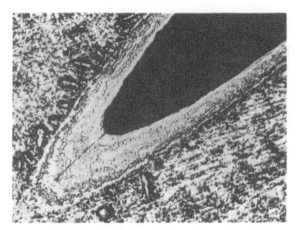

FIGURE 32. Oblique view of SiC reinforced titanium interface (500x).

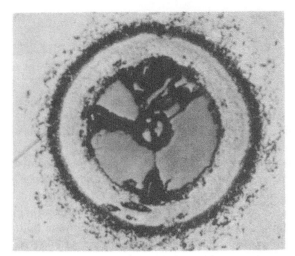

FIGURE 34. Reaction of 304 stainless steel/boron composite after 1 hr at 900°C (500x).

system, and are difficult to interpret. In Figure 34 the reaction of stainless steel with boron under the same conditions is substantially different than for iron, but the results in each case are filament destruction.[85] In Figure 35 the reaction of nickel with several filaments is observed after a 1 hr anneal (100 hr for TiB$_2$) showing further examples of multiple intermediate phase formation and destruction of the filament.[81] Similar gross reactions were found with other transition metal matrices such as cobalt and chromium.[81,85]

Snide[81] has thoroughly studied the reactions of various grades of titanium with several filaments and determined apparent rate constants based on the growth of the interaction zone. The results are

indicated schematically in Figure 36. The effects of matrix alloying in reducing the degradation of the filament are observed to be significant. The less reactive TiB$_2$ filament, unfortunately, had a relatively high 5.2 g/cm^3 density and a low average tensile strength of 140 ksi compared to other filaments such as boron. This filament apparently is no longer commercially available. The phases found at the interface between pure titanium and boron are shown as an electron micrograph in Figure 37.[81] The boron filament is in the lower portion of the figure with a distinct TiB$_2$ layer formed, and the upper layer of boron-rich

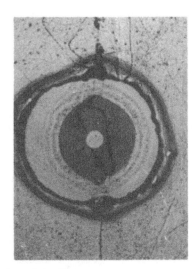
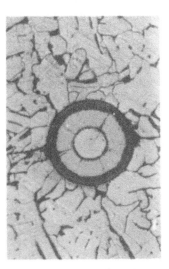

FIGURE 33. Comparison of the reaction of various filaments with iron after a 900°C anneal for various times (550x).

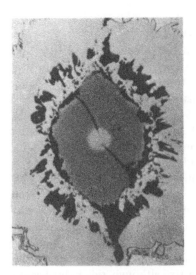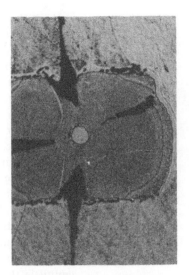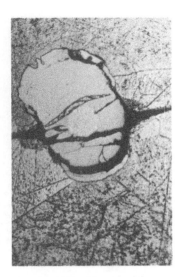

FIGURE 35. Comparison of the reaction of various filaments with nickel after a 900°C anneal for various times (550x).

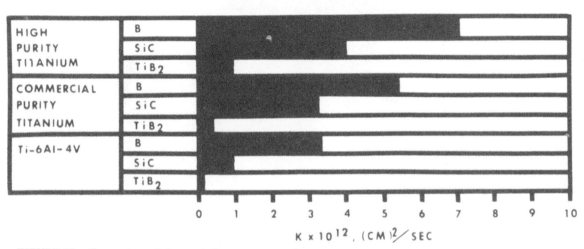

		0	1	2	3	4	5	6	7	8	9	10
HIGH PURITY TITANIUM	B											
	SiC											
	TiB$_2$											
COMMERCIAL PURITY TITANIUM	B											
	SiC											
	TiB$_2$											
Ti-6Al-4V	B											
	SiC											
	TiB$_2$											

$$K \times 10^{12}, \ (CM)^2/SEC$$

FIGURE 36. Comparison of the parabolic rate constants for the filament-matrix interaction after 100 hr at 850°C.

FIGURE 37. Electron micrograph of the reaction layer between boron and high-purity titanium after a 100-hr anneal at 850°C (1000x).

(possibly TiB) titanium moving into the matrix. The destruction of the integrity of the core of a boron-filament in titanium is shown in another electron micrograph in Figure 38. Although the tungsten core is "borided" in the as-formed filament, it is apparent here that further degradation does indeed occur under more severe environments.[81] Under similar conditions the core of a SiC filament in titanium is unchanged. The rate of reaction between a titanium alloy matrix and a boron filament can be reduced by precoating the filament with a thin titanium nitride coating as a barrier layer. The results are shown in Figure 39, where the void formation in the filament and the reaction layer width were substantially reduced

31

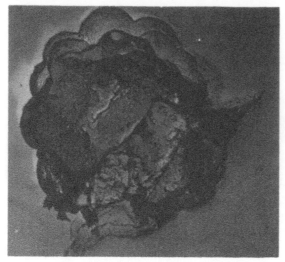

FIGURE 38. Electron micrograph of the core of the boron filament after a 100-hr anneal at 850°C (2400x).

with TiN-coated boron compared to uncoated boron in Ti-6Al-4V at 900°C.

The reaction between titanium and the boron filament proceeds with a diffusion of boron from the filament into the matrix and void formation in the filament. This is shown for titanium after 1, 9, and 100 hr at 900°C in Figure 40[81] and also observed for Ti-8Al-1Mo-1V where void formation is pronounced in 10 hours at 900°C.[85] These conditions are admittedly rigorous compared to probable use-temperatures for titanium alloys but not unreasonable in terms of consolidation conditions. The Ti-8Al-1Mo-1V alloy reaction with boron results in a rejection of aluminum adjacent to the growing TiB$_2$ layer.[8] This higher aluminum concentration in the vicinity of the interface may embrittle the matrix, although this point has not been firmly established. The reaction of silicon

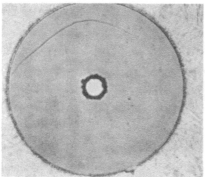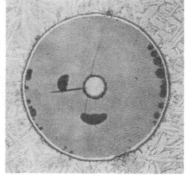

FIGURE 39. Comparison of uncoated and TiN-coated boron in Ti-6Al-4V matrix (500x (left) and 800x).

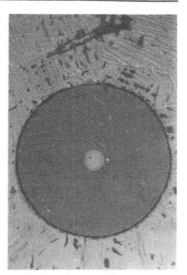

FIGURE 40. Isothermal growth of the reaction layer between the boron filament and commercial purity titanium at 900°C (550x).

carbide with titanium proceeds with the surface reaction of the filament and matrix as shown in Figure 41[70] after 1 to 100 hr at 1600°F (871°C).

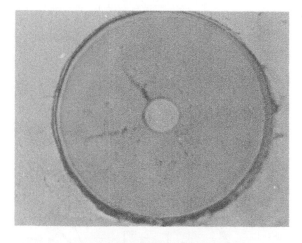

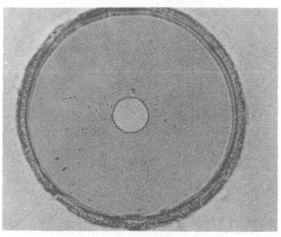

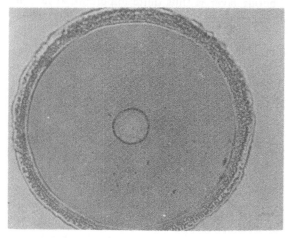

FIGURE 41. SiC-Ti interaction in Borsic-Ti composites heat treated at 1600°F: top after 1 hr; middle after 10 hr; bottom after 100 hr (700x).

Void formation is not observed and the reaction zone interface grows with the consumption of the filament. Metallographic evidence indicated the presence of complex reaction products as previously found (see Figure 32). The reaction of Ti-6Al-4V after a 100 hr anneal at 850°C is illustrated in Figure 42.[81] As with other titanium alloys, void formation was pronounced. Although this is a two-phase alloy, it was interesting to find the white α-phase completely surrounded the reaction layer at the expense of the dark β-phase. This can be rationalized as a consequence of the formation of the TiB_2 reaction layer and the rejection of aluminum ahead of the reaction front, preferentially stabilizing the phase.[81]

Controlled Interactions
Boron-Titanium

A high speed diffusion-bonding process has been employed by Metcalfe and Schmitz[71] to minimize the degradation reaction between boron filaments and titanium. Using molybdenum rolls at a temperature of 1800°F (982°C) and a contact time of about 1 sec, the reaction layer was kept to 500 Å thickness, which did not degrade composite strength. Three photomicrographs of the tape composite interface are shown in Figure 43.[70] In the narrow titanium-boron reaction layer the reaction product is predominantly titanium diboride, a hard, high modulus, brittle interlayer. Metcalfe[69] has found evidence that fracture often initiates by cracks in the diboride interlayer propagating through the filament. This is shown schematically in Figure 44. The width of the reaction layer is, therefore, of fundamental import. When the reaction product layer reaches a critical thickness, cracks in this

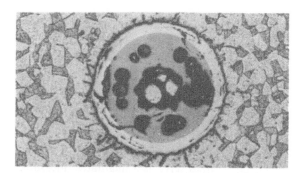

FIGURE 42. Reaction of boron filament with Ti-6Al-4V after 100 hr at 850°C (250 x).

33

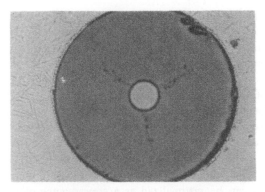

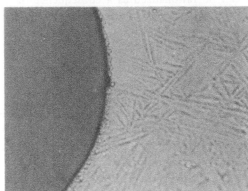

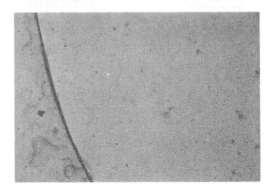

FIGURE 43. Cross section showing B-Ti interface in as-fabricated tape. (Top 700x, middle 1400x, bottom 3750x.)

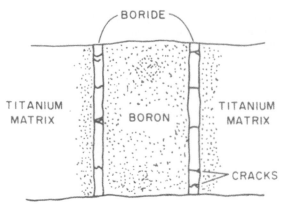

FIGURE 44. Cracks in boride layer on boron.

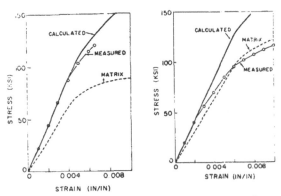

FIGURE 45. Stress strain curves for B-Ti composites.

region initiate failure in the composite. Below this thickness, pre-existing filament defects appear to initiate composite failure. Typical stress-strain curves in Figure 45 for titanium-boron composite tapes illustrate this behavior.[69] The stress-strain curve for a titanium-boron composite with a 250Å diboride thickness is seen to be close to the simple additive rule of mixtures predictions. The use of this prediction on tape materials can be questioned, but it does serve as a qualitative measure, at least of composite behavior. For a 13,000 Å-thick interaction layer the strength was found to be slightly lower than for the matrix without any filament reinforcement. In this case the filament fails at very low failure strains and the failure becomes matrix controlled. The thickness of the diboride layer is thus quite important, at least in establishing the longitudinal strength of unidirectional B/Ti composites. The reaction rates for boron and Borsic-titanium composites, expressed in reaction zone thickness, are shown in Figure 46 for the temperature range from 1200° to 1800°F (649° to 982°C). The reaction rates for Borsic decrease after the SiC-layer is consumed. An Arrhenius type plot has been employed in Figure 47 to compare data of Klein, Reid, and Metcalfe[70] with earlier work of Ashdown,[84] Snide,[81] and the earliest studies at the Air Force Materials Laboratory.[85] While this representation is helpful in qualitatively comparing the reaction rates in terms of a simple apparent "activation energy," it assumes a simplicity from a linear plot that is deceptive. For titanium-boron the reaction rates observed at 1000° to 1200°F (538° to 649°C)

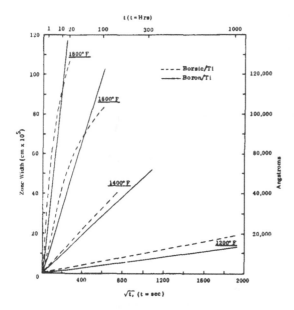

FIGURE 46. A comparison of the growth of reaction zones in B-Ti and Borsic-Ti composites.

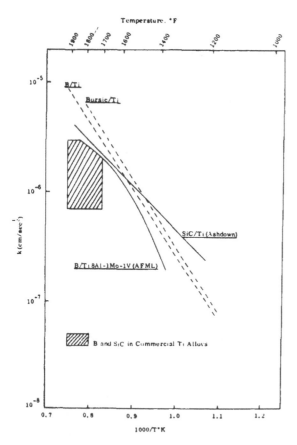

FIGURE 47. Comparison of interaction growth rates for present work with those of previous work.

were lower than the extrapolated linear plot from 1400° to 1900°F (760° to 1038°C).[70] Although the Borsic and boron reaction rates and apparent activation energies were similar, the Borsic reaction was found to be much more complex. Nevertheless, qualitative comparisons have been obtained from this approach to the data.

The longitudinal strength of unidirectional composites as a function of the interaction zone thickness for composite tapes heat-treated at 1600°F (871°C) is shown in Figure 48.[70] The initial Borsic filament strength was lower but the normalized strength (strength at time, "t"/strength at time, "t_0") was still 0.95 after 10 hr exposure which tends to correlate with the enhanced stability of Borsic filaments. From metallographic evidence the SiC coating was completely reacted after 100 hr and the strength degraded about 20%. The initial composite strength for boron-reinforced titanium was higher because of the higher initial strength of boron filaments. The longitudinal strength was found to degrade sharply when the reaction layer exceeded 4,000Å.[70] The same degradation has been observed at other temperatures (e.g., 1400°F (760°C) for longer times) when the reaction zone again exceeded 4,000Å. On this basis at 1000°F (538°C) the predicted lifetime for a titanium-boron composite system should exceed 10,000 hr without strength degradation. The transverse strength of boron-titanium was found to be insensitive to the reaction layer thickness as illustrated in Figure 49. For Borsic there was a slight decrease in transverse strength. This figure also indicates the high ratio of transverse to longitudinal strength in boron-titanium composites, compared to the rather disappointing values of 15,000 psi for transverse tensile strengths in boron-reinforced aluminum composites where the axial longitudinal strengths have routinely been 160,000 to 190,000 psi.[86] This is expected from the stronger titanium matrix.

Hamilton has also found a critical equivalent reaction layer thickness for the degradation of Borsic-Ti-6Al-4V composites to be between 1,000 and 10,000Å.[87] Some evidence of degradation of the filament strength at reaction thicknesses of 500 to 1000Å was reported, and Hamilton set diffusion-bonding parameters with 500Å thickness limit as the goal. Recently, transverse strengths approaching 90,000 psi have been achieved.[88] Tsareff, Sippel, and Herman[40] have diffusion-

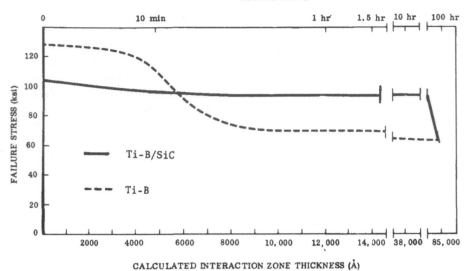

FIGURE 48. Change in longitudinal tensile strength and interaction zone thickness for composites after 1600°F exposure.

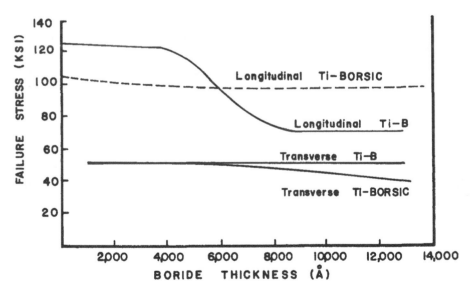

FIGURE 49. Comparison of change in longitudinal and transverse tensile strengths after 1600°F exposure.

bonded SiC/Ti6Al-4V composites at 1600°F (871°C) for 1 hr at 6,000 psi and achieved transverse strengths of nearly 100,000 psi for a composite with an axial longitudinal strength of 140,000 psi. Klein, Reid, and Metcalfe[70] have studied the reactions between a variety of transition metals commonly alloyed with titanium and many titanium alloys to determine the most stable systems with boron. Resultant reaction rate constants at 1400°F (760°C) are given in Table 3. The growth in the reaction layer with time, corrected for the initial reaction zone before heat treatment at 1400°F, is shown in Figure 50 for several compositions and alloys. From the previous data on Borsic in titanium and these data for Ti-6Al-4V an estimate of the reaction layer under the conditions given by Tsareff, Sippel, and Herman can be made, neglecting pressure effects. This

TABLE 3

Rate Constants at 1400°F (760°C) for Boron-Metal Matrix Reaction

Matrix	K (X 10^{-7} cm/sec$^{1/2}$)
Ti 40A	5.2
Ti-8Al-1MO-1V	3.4
Ti-6Al-4V	2.6
Ti-13V-11Cr-3Al (120 VCA)	1.4
Zirconium	4.2
Vanadium	2.5
Niobium	20.0
Ti-10Al	3.8
Ti-10Cu	4.7
Ti-20Sn	5.3
Ti-30Mo	1.8
Ti-30V	0.6

estimate gives a value of 6,000Å or less which is in good agreement with the observed value of 5,000Å[40] for the reaction zone. This should not substantially degrade the properties of the SiC/Ti-6Al-4V composite.

Just as a silicon carbide coating has a limited effect in improving the compatibility of titanium toward boron, a boron carbide coating appears to enhance the interactions, perhaps through promoting diffusion reactions at the matrix-filament interface.[70] Boron nitride on boron, shown in Figure 56, slightly decreases the reaction between boron and titanium.[70] The greatest effects, however, have been achieved by alloying titanium, particularly with vanadium and a high vanadium content alloy (VCA) as demonstrated in Figure 50. High molybdenum content alloys such as Beta III also have lower reactivity. Based on these results alloying has more promise than coating by an order of magnitude or more in effect on compatibility and resultant mechanical properties.

As a result of the studies of compatibility between titanium and various alloying elements with boron, two new specific titanium alloys have been developed for possible composite structures. These two alloys have reaction rates with boron that are less than one-hundredth the rate of unalloyed titanium.[89] The alloys are

a. Ti-(13-10) Ti-13V-10Mo-5Zr-2.5Al
b. Ti(22-3) Ti-22V-3W-5Zr-2.5Al

The annealed alloys are beta solid solutions with excellent ductilites, strengths exceeding 130,000 psi, and with reasonable alloy stability. They can be made into foils and the foils readily consolidated by diffusion-bonding. Problems have arisen in composite fabrication, however, indicating that considerable data on annealing and aging characteristics, high temperature properties, and stability of the alloys must be generated before successful composites can be fabricated.[89]

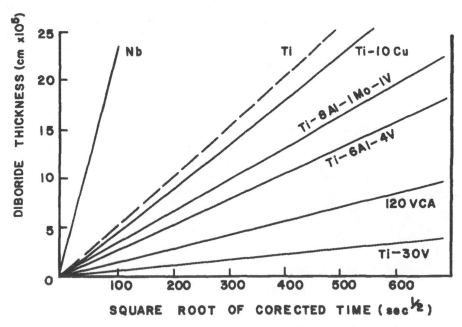

FIGURE 50. Rate of reaction between boron filaments and various metal and alloy matrices at 1400°F.

Target strengths of 150,000 psi for 25 or more vol % filament to place these composites in strong competition with (but more compatible than) B/Ti-6Al-4V have not been realized. This is quite illustrative that compatibility is only one of the major milestones that must be achieved to realize a practical composite system.

Boron-Aluminum

In subsequent sections the fabrication and variation in properties of B/Al composites with temperature are examined in considerable detail. For B/Al, reactivity in the sense of gross reactions has not been a limiting problem. Mechanical properties deteriorate with temperature sufficiently to place an upper limit optimistically of 600°F (315°C) on long-term use and 1000°F (538°C) on short-term use. These conditions are below the point of serious interaction problems. Fabrication parameters also can be adjusted to avoid conditions where measurable filament deterioration occurs. Figure 51, for example, is a photomicrograph of a B/Al composite prepared by solid state bonding at 625°C for 2 hr.[8] No evidence of reaction has been found although the bold relief of the hard boron filaments in the softer aluminum matrix gives a dark appearance around the filaments that could be mistaken for a reaction zone. One can observe how smooth a surface has been maintained on the boron filament at the interface.

The system can be pushed to gross interaction simply by employing molten aluminum (Al

melting point is 660°C) in fabrication, which also illustrates the unsuccessful use of liquid infiltration on this system as a consolidation method. Figure 52 shows the onset of reaction of a boron fiber in contact with aluminum at 1350°F (734°C) for only 3 sec.[90] The first evidences of some reaction are seen only at rather high magnification. In Figure 53 the contact time was increased to 30 sec and the result is a very definitive reaction and the onset of filament deterioration.[90]

These experiments, of course, illustrate reaction conditions that are not employed in normal fabrication of B/Al. Fairly high temperatures, however, may serve to improve the interface

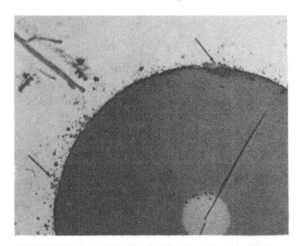

FIGURE 52. Aluminum-boron reaction product on aluminum-coated boron fiber (arrowed). This reaction product appeared tan and rather indistinct, in contrast to boron chips occasionally found in the aluminum (due to the polishing operations) which are gray and well defined (1500x).

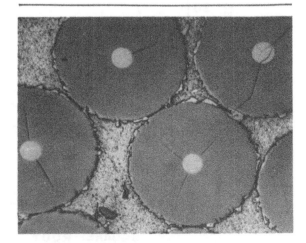

FIGURE 53. Reaction of fibers with aluminum matrix at 1350°F and 0.5 min (500x).

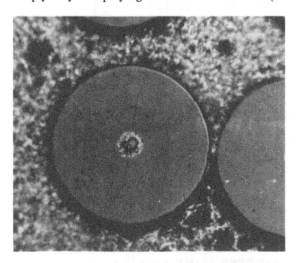

FIGURE 51. Aluminum/boron composite prepared by cold-pressing and sintering at 625°C for 2 hr (500x). No evidence of reaction.

bonding and the properties of B/Al composites. Klein and Metcalfe[91] have been investigating the effect of interface conditions on the longitudinal and transverse tensile strengths of B/Al composites by utilizing heat treatments in the range of 940° to 1100°F (504° to 593°C) for times up to 192 hr. Characterization has included metallography, scanning, and electron microscopy in the examination of the composites at the interface region, using both annealed and age-hardened alloys. While this particular effort has not been completed at the time of this writing, and the effort is continuing, several important observations have come from it already. These results indicate that improved properties for B/Al composites can be obtained through interface changes during heat treatment.

The effects of 1000°F (538°C) heat treatment on the room temperature longitudinal tensile strength of B/Al are shown in Table 4.[91] The composite was of a 48 vol % 5.6 mil boron filament in a 6061 T-6Al matrix. The strength is seen to go through a maximum after 15 min of the heat treatment and then drop sharply for several hours before leveling out at a value less than half of the original strength. In similar fashion the transverse strength of this 48B/6061 T-6Al composite was studied at 1000°F (538°C) with the results shown in Table 5. The type of filament failure has been indicated and shows correlation with interface strength. Generally, where the interface is not the predominant location of fracture, the transverse strength is relatively high. A cursory examination of the data in Table 5 reveals disturbing variations from this trend, such as the low degree of interface failure after 5- and 10-hr heat treatments where the transverse strength is low. This would seem to indicate the need for considerably more effort to examine these interface characteristics and their effect on mechanical properties. A comparison of the changes in room temperature longitudinal strength of 48B/6061 Al in the T-O and T-6 conditions is illustrated in Figure 54. In both cases a maximum in the strength is realized at very short times at the 1000°F heat treatment. On the other hand, heat treatments carried out at lower temperatures resulted in the maximum strength being reached after a longer period of time. This can be illustrated by the results shown in Figure 55, where the room temperature longitudinal strength of 48B/6061 T-OAl is seen to reach a high value in 10 min or less at 1040°F (560°C) but to take almost an hour at 940°F (504°C).[91] At 940°F the somewhat surprising result of no improvement in transverse strength as a result of the heat treat-

TABLE 4

Effect of 1000°F Heat Treatments on Longitudinal Strength
48/B/6061 T-6 Al

Time at 1000°F[1]	Failure strain (μin/in)	Elastic modulus (psi x 10^{-6})	Ultimate tensile strength	
			ksi	σ/σ_O[2]
5 min	9200	29	235	1.0
15 min	9100	31	243	1.04
30 min	8000	30	201	0.86
1.0 hr	6100	33	161	0.69
2.0 hr	5200	32	127	0.54
5.0 hr	3700	29	109	0.46
10.0 hr	3200	32	100	0.42
100.0 hr	3200	31	88	0.33

[1] After 1000°F heat treatment, specimens were given T-6 temper (quench in water and age 350°F for 7 hr).
[2] Normalized stress, σ/σ_O, is unity for zero heat treating time at 940°F.

TABLE 5

Effect of 1000°F Heat Treatments on Transverse Strength 48/6061 T-6 Al

Time at 1000°F	Failure strain (μ in.in.)	Ultimate tensile strength (ksi)	Fracture type (%)		
			Filament splitting	Matrix	Interface
0[1]	1900	20.2	15	15	70
5 min	2300	34.8	25	50	25
10 min	1770	36.8	25	60	15
15 min	2300	36.7	10	50	40
30 min	1100	33.5	20	60	20
1 hr	2700	31.2	25	40	35
2 hr	5200	20.0	40	30	30
5 hr	4600	17.1	75	—	25
10 hr	3900	20.0	15	—	85
150 hr	3100	19.0	20	—	80

[1] Zero time specimens were heat treated for 7 hr at 350°F. All other specimens were first heat treated at 1000°F, quenched in water, and aged 7 hr at 350°F.

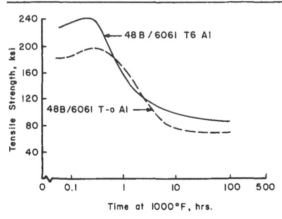

FIGURE 54. Room temperature strengths of heat treated B/Al composites.

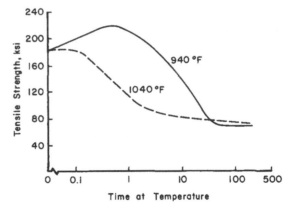

FIGURE 55. Effect of temperature on room temperature strengths of 48B/6061 T-0 Al.

ment was reported for these composites. Again there is definitely a need for further research on this problem of controlled interface reactions. Some of the results indicate the desirability of adopting post-fabrication heat treatments as a routine method for B/Al processing to optimize properties.

Alumina-Metal

The data given in Table 1 and Figure 3 indicate the attractiveness of alumina as a filamentary reinforcement to fairly high temperatures.

Moore[92] has investigated the elevated temperature compatibility of alumina filaments with nickel. In Figure 57 the formation of a reaction layer identified as the spinel $NiAl_2O_4$ by x-ray diffraction is seen in the top photomicrograph after a 100-hr vacuum anneal at 2012°F (1100°C). This reaction was shown to result from the presence of oxygen in the nickel matrix reacting to form the spinel from nickel and alumina. In this case, the reaction was extensive and filament degradation severe. By alloying the matrix with 10% Cr it was possible to preferentially bind the oxygen and

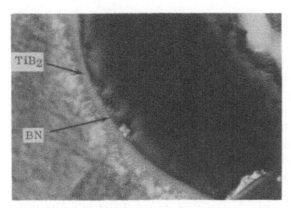

FIGURE 56. Interface of BN-coated boron in Ti 40 A matrix after 50 hr at 1400°F (1400x).

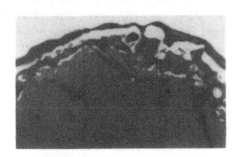

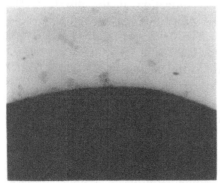

FIGURE 57. Alumina filament in pure nickel (top), showing spinel formation, and in Ni-10% Cr (bottom), showing lack of reaction (1000x).

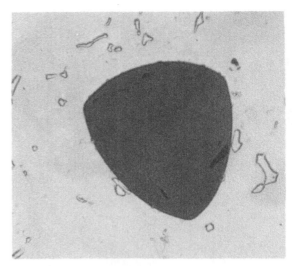

FIGURE 58. Alumina filament in Inconel 718 after 100 hr at 1100°C (200x).

virtually prevent the spinel formation. The lower photomicrograph in Figure 57 shows the interface of the alumina filament with a Ni-10% Cr matrix after the same heat treatment with no visible reaction layer. Similar results were obtained with superalloys such as Inconel 718 (in Figure 58) with no reaction with the alumina filament after

100 hr at 2012°F (1100°C). Thus, it is possible to exercise control over the spinel reaction through alloying. Another problem at higher temperatures was the degradation of the alumina filament surface without the appearance of a distinguishable third phase.[92] This, apparently, is a thermal etching attack which is being investigated by Mehan for nickel and Ni-Cr matrices.[93] Moore observed a resultant loss of strength in titanium and nickel coated alumina filaments at higher temperatures, which would suggest a limiting compatibility range without third phase formation. A scanning electron micrograph of a composite at fracture in Figure 59 shows that the debonding between nickel and alumina occurs where a sufficient bond strength has not been developed through heat treatment.

In further studies of alumina filament transition metal composites, Tressler and Moore[94] have found that neither nickel nor niobium forms chemical bonds sufficient to eliminate debonding. For nickel, complete mechanical consolidation was effected without obtaining adhesion of filament and matrix. The fabrication and mechanical properties of alumina-titanium composites have been investigated by Tressler and Moore.[95] Interface reaction kinetics were treated in the manner of previous investigators[4,5,70,81,84,85] and Ti_3Al was identified by electron microprobe analysis as a principal reaction product. An important feature is shown in Figure 60, where an α-titanium phase ring was formed around an alumina filament after

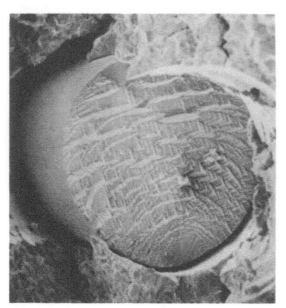

FIGURE 59. Scanning electron micrograph of alumina filament in nickel (500x).

FIGURE 60. Alumina filament in titanium after 2 hr at 800°C (200x).

2 hr at 1472°F (800°C). The reaction zone is also clearly visible. The growth of this zone was slightly faster with alumina filaments than for titanium-boron reactions in Ti-40A and Ti-6Al-4V matrices as listed in Table 2.[70] The tensile strength of boron-titanium is much more sensitive to chemical degradation than alumina-titanium. Alumina filaments, uncoated, actually increase in strength[95] after a 1400°F (760°C) heat treatment, probably through surface annealing and residual stress relief. Filaments coated with titanium decreased moderately in strength after 100 hr at 1400°F (760°C).[92] At this point boron filaments have substantially lost strength. Heat treatment of alumina-titanium composites at 1500°F (816°C) for periods up to 8 hr actually increased longitudinal and transverse strength although longer heat treatments then degraded strength. This may indicate that a limited amount of chemical interaction optimized bonding. Microhardness tests after 69 hr at 1500°F indicated substantial matrix embrittlement from dissolved oxygen from the filament. The strength was also degraded, particularly in transverse specimens which are more sensitive to matrix properties. Matrix embrittlement in the Borsic titanium composite is believed to be due to the analogous carbon embrittlement.[83] Interesting, too, was the increased reaction rate of alumina filaments with Ti-6Al-4V compared with the alumina/Ti-40A reaction from 1200° to 1600°F (649° to 871°C). In this system, alloying enhanced the reaction probably through the influence of Al in the matrix on Ti_3Al formation at the interface. The preliminary properties reported for a 22 vol % alumina/Ti-6Al-4V composite were 125,000 psi longitudinal and 57,000 transverse tensile strengths with moduli of 27 and 17 x 10^6 psi, respectively.[95]

In a later report by Tressler and Moore[96] the composite tensile data have been listed as given in Table 6. Apparently, no improvement in properties beyond those initially reported has been achieved. Filament misalignment and incomplete consolidation resulting in a poor interface bond continue to limit composite performance. The longitudinal strengths realized with Al_2O_3/Ti composites have not been competitive with the B-SiC/Ti competitor materials. The transverse strengths, however, have been attractive although the use of wide diameter boron with its improved transverse properties will probably remove this advantage as well. It is instructive to consider further the effects of alloying on the Al_2O_3/Ti reactions at the interface. The graphical results of interaction zone thickness with time at temperature are shown for two titanium alloys in Figure

TABLE 6

Al_2O_3/Ti-6Al-4V Composite Tensile Data

Sample	Orientation of filament to test axis	Vol fraction Al_2O_3 %	Failure strain, microin./in.	Elastic modulus, 10^6 psi	Ultimate strength, 10^3 psi	Remarks
1	0°	28	5000	22	108	
2	0°	22	5000	21	106	
3	0°	22	5600	27	125	
4	0°	23	4600	21	119	
5	0°	20	5400	21	118	
6	0°	20	13000	22/19	123	Poor bonding. Primary and secondary modulus.
7	0°	28	5300	27/19	91	Grooved foil. Poor bonding. Primary and secondary modulus.
8	0°	22	2000	28/19	98	Poor bonding. Primary and secondary modulus.
9	0°	20	5500	24	98	Poor bonding.
10	0°	28	5800	27	71	Poor bonding.
11	90°	22	2800	17	40	Poor bonding. Grip failure.
12	90°	22	No data	No data	57	Well bonded.
13	0°	22	12000	22	99	As fabricated.
14	0°	22	12000	22	105	Annealed 8 hr at 1500 F.
15	0°	22	11500	24	96	Annealed 69 hr at 1500 F.
16	90°	22	6600	17	51.8	As fabricated.
17	90°	22	6200	17	52.6	Annealed 8 hr at 1500 F.
18	90°	22	2200	24	24.3	Annealed 69 hr at 1500 F.

61.[29,96,97] The reaction rates for Ti-8Al-1V-1Mo are much higher than for Ti-6Al-2Sn-4Mo-2Zr which conforms in part, perhaps, to the view that the higher amount of aluminum in the matrix enhances the formation of Ti_3Al at the interface. Perhaps more significant is the effect of alloying elements such as molybdenum and zirconium on lowering the reaction rate, as has been observed in the B/Ti system. In this example, the rate of reaction of the Ti-8Al-1V-1Mo alloy at 1450°F (787°C) is approximately the same as the reaction rate of the Ti-6Al-2Sn-4Mo-2Zr at 1600°F (871°)C. The work of Tressler and Moore[96] is particularly notable in the excellent use that was made of scanning electron microscopy to show the failure surfaces of the Al_2O_3/Ti composites and to determine the nature of the defects apparently responsible for premature failure in some composite specimens.

Thermodynamic considerations of the free

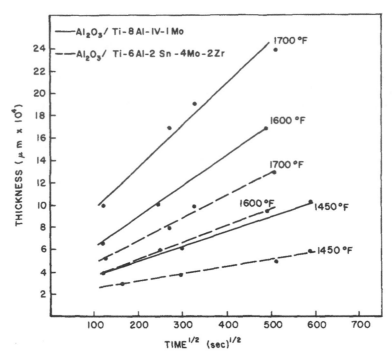

FIGURE 61. Growth of reaction zone thickness as a function of time and temperature.

energy changes (ΔF) for possible reactions provide a starting point for the selection of Al_2O_3/M systems stable at elevated temperatures.[4] In Figure 62, ΔF for a series of reactions of possible matrix metals with continuous alumina filaments (to reduce the alumina and thus degrade the filament properties) is given as a function of temperature.[98] Those metals which readily reduce alumina have ΔF values that are negative or close to zero. Nickel, for example, has a high positive value for ΔF, indicating considerable thermal stability since the reaction

$$Ni + 1/3\ Al_2O_3 \rightleftharpoons NiO + 2/3\ Al$$

lies strongly to the left. One of the early whisker reinforced composite systems investigated was the Al_2O_3/Ni system because of its inherent stability.[99] As Moore has shown,[92] the formation of $NiAl_2O_4$ at the interface shows one of the limitations of this approach. In like manner the reaction of Ti and Al_2O_3 shown as

$$Ti + 1/3\ Al_2O_3 \rightleftharpoons TiO + 2/3\ Al$$

would not account for the formation of Ti_3Al which has been identified as a major reaction product.[96]

A TiO phase has also been identified in the reaction zone immediately adjacent to the alumina filament after an 1800°F (981°C) heat treatment for 940 hr.[97] So the interface reactions are reasonably complex. Despite the limitations of only considering the fundamental Al_2O_3/M reactions for the reduction of the alumina filament in the filament degradation process, the ΔF curves of Figure 62 provide a useful starting point for selecting stable systems and considering possible alloying elements for various metal matrices.

Mehan and Harris[100] have investigated the preparation and properties of Al_2O_3/Ni-Cr composites and the interface reactions of Al_2O_3 filaments with nickel, nickel alloys, and particularly the Ni-Cr alloys. In the course of this study a quantitative treatment of the Al_2O_3/Ni interface reaction to form the $NiAl_2O_4$ spinel was given. Using metallographic taper sections to determine the reaction zone thickness, the data shown in Figure 63 were obtained. While the method of establishing a reaction rate is not highly accurate, it does provide a comparison to the Al_2O_3/Ti system. In this temperature range the reaction rate for Al_2O_3/Ti is apparently at least tenfold greater than for Al_2O_3/Ni. The Al_2O_3/Ni-20Cr system was also studied after heat treatment at 1200°C

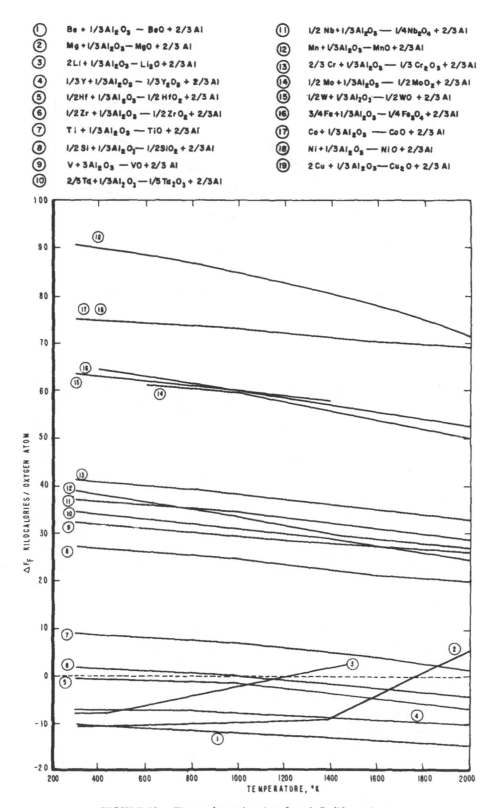

① Be + 1/3 Al₂O₃ — BeO + 2/3 Al
② Mg + 1/3 Al₂O₃ — MgO + 2/3 Al
③ 2 Li + 1/3 Al₂O₃ — Li₂O + 2/3 Al
④ 1/3 Y + 1/3 Al₂O₃ — 1/3 Y₂O₃ + 2/3 Al
⑤ 1/2 Hf + 1/3 Al₂O₃ — 1/2 HfO₂ + 2/3 Al
⑥ 1/2 Zr + 1/3 Al₂O₃ — 1/2 ZrO₂ + 2/3 Al
⑦ Ti + 1/3 Al₂O₃ — TiO + 2/3 Al
⑧ 1/2 Si + 1/3 Al₂O₃ — 1/2 SiO₂ + 2/3 Al
⑨ V + 3 Al₂O₃ — VO + 2/3 Al
⑩ 2/3 Ta + 1/3 Al₂O₃ — 1/5 Ta₂O₃ + 2/3 Al

⑪ 1/2 Nb + 1/3 Al₂O₃ — 1/4 Nb₂O₄ + 2/3 Al
⑫ Mn + 1/3 Al₂O₃ — MnO + 2/3 Al
⑬ 2/3 Cr + 1/3 Al₂O₃ — 1/3 Cr₂O₃ + 2/3 Al
⑭ 1/2 Mo + 1/3 Al₂O₃ — 1/2 MoO₂ + 2/3 Al
⑮ 1/2 W + 1/3 Al₂O₃ — 1/2 WO + 2/3 Al
⑯ 3/4 Fe + 1/3 Al₂O₃ — 1/4 Fe₃O₄ + 2/3 Al
⑰ Co + 1/3 Al₂O₃ — CoO + 2/3 Al
⑱ Ni + 1/3 Al₂O₃ — NiO + 2/3 Al
⑲ 2 Cu + 1/3 Al₂O₃ — Cu₂O + 2/3 Al

FIGURE 62.   Thermodynamic values for Al₂O₃/M reactions.

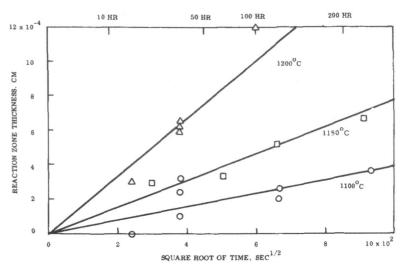

FIGURE 63.   Growth of Ni Al$_2$O$_4$ in Ni/Al$_2$O$_3$ composites.

for 200 hr. X-ray studies showed evidence of some NiO and NiAl$_2$O$_4$ but no Cr$_2$O$_3$ formation. Very little reaction is found, however, and the conditions of a 1200°C heat treatment represent a maximum that one might expect in using a Ni-Cr base composite, since the matrix strength is less than 10,000 psi at this point.

Fabrication parameters were established by Mehan and Harris[100] for the Al$_2$O$_3$/Ni-Cr composites using Y$_2$O$_3$-coated punches and die parts to prevent reaction between the composite specimens and the hot press. Preferred conditions were 1100°C for 1 hr at 5,000 to 6,000 psi in a vacuum of 5 x 10^{-5} mm Hg. The results of these experiments are given in Table 7. The experimental values for strengths were quite disappointing. Changes in the fabrication parameters did not improve the results and considerable filament damage was noted even at relatively low volume loadings. At higher volume loadings (25 to 30% alumina filament) extensive fiber fracture occurred, even under conditions so moderate that the consolidation was poor, such as 4,250 psi at 1000°C. This is indicative of the need for a filament with higher compressive strengths to withstand the processing conditions. Attempts were made to improve the bond strength by sputtering a thin (0.5μ) titanium coating on the alumina filament. It has been previously shown by Sutton and Feingold[101] that interfacially active metals like titanium would migrate to the alumina-metal interface and react to form new phases. Using the transverse strength as the measure of

improvement in bond strength, no significant change was found by sputtering a titanium or a nickel-chromium coating on the alumina filament.

Mehan and Harris[100] utilized the predictions of Chen and Lin[102] and Cooper and Kelly[103] to determine the effectiveness of obtaining a good bond in the Al$_2$O$_3$/Ni-Cr composites. These results are indicated in Figure 64, which shows that the values for the transverse strength at various volume loadings of filament closely correspond to a no-bond situation. For this type of composite there appears to be a need for a great deal of further effort before it will have attractive properties as a high temperature material.

Several recent fundamental studies in an Al$_2$O$_3$/M system may be useful in providing the background necessary to establsih good bonding. Nicholas[104] studied the strength of a number of alumina-metal interfaces. Correlations between physical properties and interface strengths were not obtained. For a wide variety of metals reasonably good interfacial strengths were found. The conditions for obtaining a strong bond between nickel and alumina have been examined in some detail with nickel powder specimens hot-pressed to alpha-alumina single crystal disks.[105] The interfacial shear strength reached a maximum at temperatures of 900 to 1000°C at pressures of 7,000 to 8,000 psi. Pressing time was optimized at 1 hr or more than 24 hr for the 1000°C hot pressing. The bonding of nickel to alternate sheets of nickel and chromium and nickel-chrome to alpha-alumina has also been

TABLE 7

Summary of Al$_2$O$_3$/Ni-Cr Results

Sample	Orientation of filament to test axis	Matrix metal	Filament diameter inches	Coating	Heat treatment	Volume fraction Al$_2$O$_3$ %	Composite strength psi	Ratio: composite strength matrix strength
1	0°	Ni	—	—	—	—	41,500	—
2	0°	Ni-Cr	—	—	—	—	81,500	—
3	90°	Ni-Cr	0.010	None	None	15.6	35,900	0.44
4	90°	Ni-Cr	0.010	None	1100°C-120 hr	~16.0	37,000	0.45
5	90°	Ni-Cr	0.010	None	None	—	32,600	0.40 [a]
6	90°	Ni-Cr	0.010	None	None	—	36,400	0.45
7	90°	Ni-Cr	0.010	Ni Cr	None	21.6	—	—
8	0°	Ni-Cr	0.010	None	None	9.5	—	—
9	0°	Ni-Cr	0.010	Pt-Pt/Rh	None	~10.0	55,000	—
10	90°	Ni-Cr	0.010	None	None	—	28,700	0.352 [a]
11	0°	Ni-Cr	0.010	None	None	—	—	—
12	90°	Ni-Cr	0.010	None	1200°C-48 hr	17.4	26,400	0.324 [a]
13	0°	Ni-Cr	0.010	Ni Cr	None	6.5	76,500	—[b]
14	90°	Ni-Cr	0.010	Ni Cr	1200°C-48 hr	14.0	42,000	0.515
15	90°	Ni-Cr	0.020	Ni Cr	None	35.7	34,000	0.295
16	0°	Ni-Cr	0.020	Ni Cr	None	13.9	51,700	—[c]
17	90°	Ni-Cr	0.020	Ni Cr	None	38.9	19,200	0.236
18	90°	Ni-Cr	0.010	Ti	None	31.5	28,200	0.346
19	90°	Ni-Cr	0.010	Ti	1200°C-48 hr	20.0	39,600	0.486
20	0°	Ni-Cr	0.010	None	None	18.0	84,600	—[d]

[a] Composite misalignment of filaments
[b] 324,000 psi fiber stress determined from strain at the first fiber break
[c] 223,000 psi fiber stress determined from strain at the first fiber break
[d] 31 x 10^6 psi elastic modulus

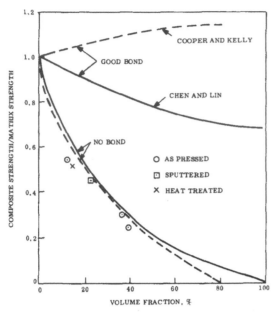

FIGURE 64. Transverse composite strength.

studied by the same hot pressing techniques.[106] The optimum conditions varied considerably depending on the order of the layers of metals between alumina disks, with higher temperatures up to 1300°C giving higher strengths. Commercial Ni-Cr containing approximately 1% silicon gave poorer results compared to high purity material.

Graphite-Metal

The potential for graphite in metal matrices has been studied by Jackson.[20] Both chemical reaction and structural recrystallization were observed when graphite fibers were coated with various metals such as aluminum, nickel, cobalt, copper, platinum, and nickel-chromium. The unstable carbide Al_4C_3 was found, for example, after heat treatment of aluminum-coated graphite fibers at 600°C. Structural recrystallization was observed extensively with nickel-graphite at 1000°C with a marked loss in strength of the graphite at 900°C. For aluminum-graphite, a sharp strength drop-off occurred at 600°C, thus limiting potential processing conditions. Chromium readily formed the carbide Cr_3C_2 at 1000°C, weakening the structure, and the fiber strength fell markedly at 600°C. Graphite fiber strengths decreased in the 600°C range for cobalt, copper, and even platinum, indicating a general limitation for at least the current generation of very fine diameter graphite fibers. Coatings for minimizing reactions

must be considered for most systems. Burte and Lynch[4] previously found that few metal candidate matrix materials do not react deleteriously with graphite at elevated temperatures. Meiser and Davison, however, have achieved stability of graphite fibers in an iron-aluminum alloy which has a low solubility for graphite.[107] An example of the stability of a graphite fiber in an Fe-Al alloy at 900°C is shown in Figure 65. Without a high strength, high modulus graphite *monofilament*, Meiser and Davison[107] point out that difficulties in fabrication preclude the successful application of graphite in metal matrices.

If the researcher limits himself only to the compatible systems that do not readily react with carbon, he is indeed limited in choice. This is illustrated in Table 8, where the candidate metals are listed with their eutectic data and melting points.[4] The cobalt and nickel matrices may appear attractive, but they exhibit a high solubility of graphite at relatively low temperatures. Metals like copper and compounds like Cr_3C_2 which have low rates of diffusion for carbon have potential use as diffusion barriers between the matrix and the filament. Other compounds such as B_4C, which has been demonstrated to protect boron filaments against attack in a nickel matrix at 800°C for 24 hr,[108] might be considered for graphite. Metals such as platinum, rhodium, and

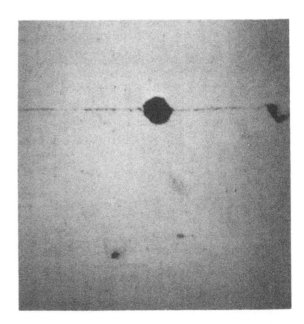

FIGURE 65. Graphite fiber in Fe-6Al alloy after 40 hr at 900°C (750x).

TABLE 8

Selected Metals That Are Not Carbide Formers

Metal	Melting point °C	Eutectic point °C
Re	3180	2486
Os	3000	2732
Ru	2500	1942
Ir	2440	2296
Rh	1966	1694
Pt	1773	1736
Pd	1550	1504
Co	1495	1309
Ni	1452	1318
Cu	1083	none

iridium would seem to involve prohibitive costs in any practical systems.

A detailed account of the graphite filaments in the yarn form as commercially available is given by Galasso.[24] Recently Hough[109] has vapor deposited a carbon fiber on a carbon substrate by chemical vapor deposition, resulting in a carbon monofilament with a diameter of 3.5 mils. Strengths up to 465,000 psi and elastic moduli to 28×10^6 psi have been reported. The deposition atmosphere contained hydrogen, argon, ethylene, and triethyl borane which resulted in a mixed boron-carbon deposition where the outer deposition layer contained a high percentage of boron. In a matrix compatible with boron, such as aluminum, this is no problem. For many matrices, the outer layer of boron would only enhance filament degradation and be quite unsatisfactory. Substantial property ranges have also been noted with this type filament, but it must be considered as a promising advance toward producing a stable monofilament in the 3 to 4 mil range. Both Hough[109] and Ezekiel[110] in reviewing the status of graphite fiber technology, have shown that the strength of graphite fibers drops rapidly when diameters exceed the range of 10 to 15μ (or approximately 0.5 mil). It does not matter what the source of fiber is, for example, from cellulosic yarns, acrylonitrile base polymers, aromatic polyamides, polyvinyl alcohol, pitches, etc., the results are essentially the same and no strong wide diameter monofilament graphite of high strength and modulus has been marketed. This also includes attempts to build up pyrolytic carbon on tungsten.

Another recent approach to large diameter filaments of graphite and carbon was to impregnate graphite fiber bundles with a high-char polymer and then convert the polymer to carbon by pyrolyzation.[111] The best results were obtained using a polyacrylonitrile precursor tow and a furfuryl alcohol binder, formed by a molding process. The impregnated fiber bundles were placed in a grooved steel base plate, covered with a soft aluminum wire placed over the molding channel, and then by a steel top plate. The fiber bundles were then hot-pressed and cured. The pyrolyzed composite graphite/carbon fiber had a nominal diameter of 8 mils, with 76 vol % fiber, and best properties of 175,000 psi tensile strength and modulus values exceeding 50×10^6 psi. Problems encountered in the process resulted in nonuniform distribution of fibers and porosity, cracking within the pyrolyzed carbon matrix, and irregular cross-sections on the relatively short lengths (3 in.). Attempts to utilize processes amenable to production of long continuous filaments by rolling and drawing techniques were not very successful.

As mentioned previously, with the lack of large diameter monofilament graphite fibers, some investigators have chosen to proceed with the preparation of metal composites utilizing the graphite yarns that are easily obtained. A substantial amount of research has been conducted on graphite/Ni composites using the "Thornel®" fiber.[112] The results have not been encouraging. To improve the processing procedures nickel was electroplated on the yarn. The nickel-coated yarn was then diffusion-bonded to the nickel matrix by vacuum hot pressing. Typical conditions were 1050°C for 1 hr at 2,000 to 3,500 psi and vacuums to 1×10^{-5} mm Hg in a graphite die. A lower fabrication temperature was achieved by employing preform tapes with alternating layers of nickel. Good densification using the tapes was achieved at 700°C and 3,000 psi, but the properties were inferior to those of the electroplated yarn composites. A great deal of trouble was experienced in trying to obtain adequate bonding and consolidation in these composites.[112] This is reflected in Figure 66, where the strength of the graphite/Ni system is shown from room temperature to 1050°C. The "Thornel" 75 is seen to be superior to "Thornel" 50 graphite fiber, but in either case the strength falls rapidly at temperatures above 400°C.

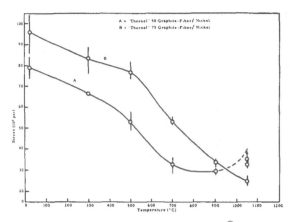

FIGURE 66. Tensile strength of "Thornel®" 50 and "Thornel®" 75 graphite-fiber. Nickel-matrix composites, 45 vol % fiber.

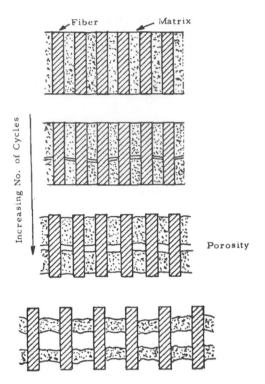

FIGURE 67. Composite degradation from thermal cycling.

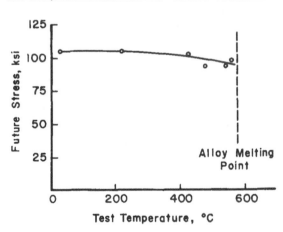

FIGURE 68. Temperature dependence of fracture strength.

The flexural strength of the "Thornel" 75 graphite/Ni composites was observed to drop very sharply above 350°C. Other properties also were not adequate, such as a stress-rupture value of about 50,000 psi for 1 hr and over 40,000 psi for 100 hr at 500°C. Since the major interest in graphite/Ni should be at higher temperatures, these data are unsatisfactory. Even more negative are the results of thermal cycling of these composites. The substantial thermal expansion mismatch between graphite fibers and the nickel matrix, plus the lack of an adequate interfacial bond leads to complete failure in thermal fatigue. Cycling at 125 to 500°C for 325 cycles, for example, leads to delamination and complete pull-away of the nickel from the fibers, as depicted in Figure 67. This also is accompanied by extensive cracking and a complete loss of integrity of the composite structure.[112]

Several reports have been published on the preparation of a graphite yarn reinforced aluminum.[113-115] A chemical pretreatment of the "Thornel" 50 yarn was devised to completely remove surface contaminants from the interior of fiber bundles as well as the surface. Many pertinent details are missing, but the preliminary reports indicate that the pretreated yarn is then infiltrated with a low-melting aluminum alloy, typically Al-13Si, and then hot formed with the molten aluminum alloy to make test specimens. The results of testing a 28 vol % graphite/Al-13Si composite are shown in Figure 68 from room temperature virtually to the alloy melting point.[113] The results show excellent strength retention with temperature and a composite strength equal to that predicted by a rule of mixtures calculation. Actually the values given for tensile strength vary from 65,000 psi to 144,000 psi with an average of 106,000 psi.[115] This indicates a need for substantially more test data on a greater number of specimens. The transverse properties should also be examined. Some evidence of the carbide phase, Al_4C_3, was reported[113] which could be expected to create problems in service by reacting with water vapor. Apparently, this has not been a problem during this preliminary work.

Fabrication Effects

The work of Stuhrke and co-workers[79,116,117] on the fabrication of B/Al is illustrative of the importance of bonding conditions on resultant mechanical properties. The composite microstructure shown in Figure 69 depicts a consolidated specimen where the aluminum foil boundaries remain and the bonding is incomplete although there is apparently good filament-foil contact without severe filament degradation. Varying the fabrication parameters eliminated the aluminum-aluminum foil interface while maintaining the filament intact, as shown in Figure 70. Processing conditions have been investigated to maximize the strength and stiffness properties for B/Al composites.[79] The optimized processing conditions found for diffusion-bonded B/Al are given in Figure 71. Stuhrke was able to achieve a filament-matrix bond with sufficient strength to experimentally demonstrate a positive deviation from a rule-of-mixtures prediction. This "synergistic" effect was predicted by Ebert and

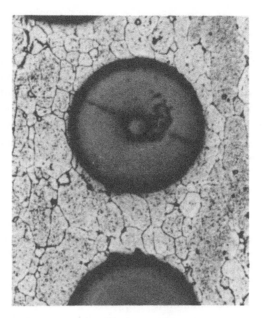

FIGURE 70. Boron-aluminum composite showing elimination of Al-Al foil interface (complete bond) (500x).

Gadd[77] and Ebert, Hamilton, and Hecker[78] and arises from the multiaxial stress state in the matrix. This could also have been predicted from the work of Shaw, Shepard, and Wulff.[118] Stuhrke[119] has recently observed that after heat

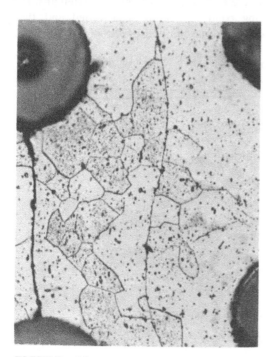

FIGURE 69. Boron-aluminum composite showing incomplete Al-Al bond (400x).

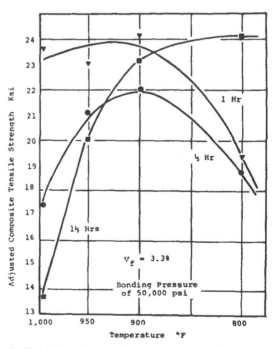

FIGURE 71. Composite longitudinal tensile strength as a function of bonding conditions.

treatment at temperatures to 450°F (232°C) this synergistic behavior presisted to 1,000 hr but was lost at 5,000 hr with a consequent unexplained reduction in composite tensile strength. At 700°F (371°C) the "synergistic life" was substantially reduced with a steady decrease in tensile strength with time. At 1000°F (538°C) the "synergistic life" was found to be less than 1 hr and the composite strength degraded rapidly for 10 hr and then more slowly thereafter. After 1,000 hr at 1000°F, however, the tensile strength was still in excess of 40,000 psi, which was much higher than the unreinforced 1145 aluminum matrix. No observable reaction or change in grain structure was observed by Stuhrke in these heat treatments.[119] Previous work by Clark,[120] however, indicated that the diffusion of boron into aluminum to a depth of 6μ from vacuum deposited aluminum or boron would produce mechanical properties equivalent to Stuhrke's 1,000-hr treatment at 1000°F (538°C). Hill and Stuhrke[121] have found interaction products at higher temperatures, moreover, in studying the preparation of cast boron-aluminum composites using liquid infiltration. In this case, gross filament degradation resulted in a very low strength composite. The decline with time of room temperature composite tensile strength after heat treatments at several temperatures has been compared for boron reinforced aluminum and titanium in Figure 72. The tensile strengths have been normal-ized to the original values for as-fabricated composites. The titanium composites have much greater potential at 1000°F (538°C) as found in the negligible decline in normalized strength which is less than 3% after 5,000 hr. The data in Figure 72 for the Stuhrke[119] results at 450°F (232°C) have been plotted in linear fashion for the entire curve; but actually, the values do not drop until beyond 100 hr ($h^{1/2} = 10$) and insignificantly until beyond 900 hr ($h^{1/2} = 30$). The tensile modulus values do not show this plateau effect but decrease rapidly at all temperatures above 450°F (232°C) in 100 hr or less. The Solar data, as discussed, relate to the formation of a reaction product layer.

Toth[122] has found that the transverse tensile fracture of B/6061 Al composites in the annealed condition for 30 to 70 vol % boron was controlled by the matrix strength of the weak matrix. For stronger matrices (through solution treating and aging) failure was controlled by the transverse strength of the filaments.

Kreider, Dardi, and Prewo[123] have found fiber splitting to be a limiting factor in transverse strengths. They have employed heat treating of Borsic/2024 Al composites to increase matrix strengths while producing a residual compressive stress on the fiber to inhibit splitting. This has substantially increased the observed strengths. Hot pressing conditions have been varied and their effect on Borsic composites with 6061 Al, 5052

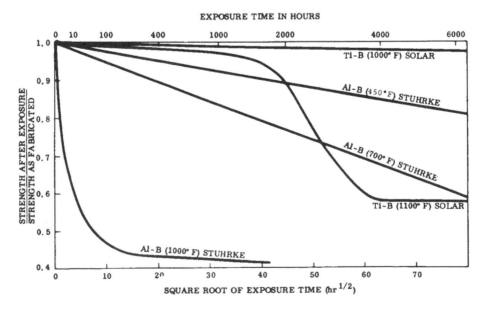

FIGURE 72.   Degradation of composite room temperature strength thermal exposure.

Al, and Al-7% Si determined. The transverse moduli increased at pressures of 2,000 to 10,000 psi and at temperatures from 420°C to 565°C for all composites except Al-7% Si where the curve did not change with pressure at 565°C. The maximum increases ranged from one to three times the low values and the curves tended to flatten out under the more rigorous conditions. The transverse strengths were less sensitive to hot pressing conditions and for 2024 Al a temperature of 490°C was found to degrade the strength with increasing pressure. Using one- and four-hour times for hot pressing did not make much difference in measured values for strength or modulus. Comparing these results to those of Stuhrke[79,119] in Figures 71 and 72 it is noted that the transverse properties are less sensitive than the longitudinal properties to the length of time for high temperature exposure. This observation is consistent with the filament-splitting concept as a limiting factor for transverse loading.

Antony and Chang[124] have conducted a comprehensive study of the properties of diffusion-bonded B/Al composites. Their "unexpectedly low" room temperature tensile strengths probably result, however, from the bonding conditions of 900°F (482°C) at 4,000 psi for 30 min. More rigorous optimum conditions are indicated by other investigators.[79,122]

Breinan and Kreider[37] have utilized braze-bonding for consolidation of B/Al composites at pressures of only 15 to 200 psi. This important effort produced composites from plasma-spray monolayer tapes of Borsic/Al with a 713 aluminum brazing foil with tensile strengths of up to 160,000 psi for a 52 vol % boron composite.

Cold working of B/Al composites by transverse rolling (90° to the direction of reinforcement) has been studied by Getten and Ebert.[125] This deformation had a beneficial effect in raising the tensile strength of the composites and increasing the strain range over which the totally elastic strain response was obtained. The results were attributed to strain hardening of the matrix and mechanical relief of residual stresses in the as-fabricated composites. The cold working effect was proportionately greater with the higher vol % boron fiber. For a 44.5 vol % boron composite, the maximum cold working practical was 30%. Beyond this point fiber breakage became significant and the strength deteriorated rapidly. The improvement in strength was larger than could be attributed to cold-working effects in the matrix alone. The composite moduli did not change with cold working, but the fracture strain was increased slightly.

Hecker, Hamilton, and Ebert[126] have studied the effect of residual stresses on composite tensile behavior of tungsten-reinforced copper composites and the use of axial straining for mechanical residual stress relief. Fabrication-induced residual stresses were found to be detrimental to the composite tensile behavior. A prestrain of 0.3% for W/Cu composite specimens caused a small permanent dimensional change which led to a 20% improvement in subsequent tensile performance.

The evidence available indicates the value of optimizing initial fabrication procedures and secondary processing steps. As the field of metal matrix composites matures this will undoubtedly be an area of expanded interest.

Chapter 5

REFRACTORY METAL WIRE COMPOSITES

Although the major use of refractory metal wires has been as substrates for other reinforcement materials like boron, substantial interest has been shown in their utilization as high temperature reinforcements as such. Table 1 shows the attractive strengths and moduli for molybdenum and tungsten, for example. Because of its ready availability and relatively low cost as lamp wire, tungsten has provided a model system material for most studies undertaken. At lower temperatures (below 500°C) the higher specific strengths of a boron (on tungsten) filament with a strength to density ratio of 4.68 compared to 0.83 for tungsten preclude serious interest in refractory wire reinforcements. At higher temperatures for such applications as the reinforcement of super-alloy matrices, the properties of tungsten become attractive despite the high density, as indicated in the data shown in Figure 3.

Weeton and Signorelli[127,128] have reviewed the earlier attempts to develop ceramic and metallic filament reinforced refractory metal and superalloy composites and concluded that a significant improvement in stress-rupture properties at temperatures of about 2000°F (1093°C) should be possible utilizing high strength metal fibers. Petrasek and Signorelli[129] have obtained reasonable properties in a study of tungsten alloy fiber reinforced nickel base alloy composites. Such composites combine the high strength of the refractory metal fibers with the oxidation resistance of superalloy matrices for possible improved turbine materials. Some of the properties and difficulties associated with the development of these type materials and some of the important model systems behavior studies will be presented in this section.

Tungsten-Copper Model System

The feasibility of refractory wire reinforcement was initially demonstrated by McDanels, Jech, and Weeton.[130,131] The tungsten-copper system was particularly suitable because of the higher strength and modulus of tungsten, the low mutual solubility of tungsten and copper, and the ability of copper to wet tungsten. Composites were prepared by liquid infiltration of bundles of axially aligned tungsten wires in a ceramic tube with molten copper. Increasing the vol % of tungsten in the copper matrix indicated that the strength of both continuous and discontinuous fiber composites varied directly with the volume of the reinforcement.

The early advantages of continuous filaments over discontinuous filaments that were shown in these studies should have been sufficient to shift emphasis in all composite studies to the continuous filament reinforcements. Unfortunately, the shift away from discontinuous filaments and whiskers did not occur until the late 1960's although the evidence was accumulating for almost five years. We have included a later section on fine filamentary reinforcements with the reservation that economic production of the reinforcements must precede any large-scale development.

In 1967 Petrasek, Signorelli, and Weeton[132] pointed out that the orientation of the fibers in a discontinuous fiber reinforced composite was the "most significant factor in achieving higher tensile strength at elevated temperatures." This is assuming a critical aspect ratio (length to diameter L/D) is maintained at 100 or greater during all fabrication and subsequent processing steps. The importance of the L/D ratio is indicated in Figure 73. The stress required to break a tungsten wire or

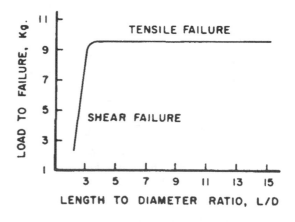

FIGURE 73. Experimental length to diameter ratios for tungsten fibers in copper at room temperature.

pull it from the copper wire was determined by Jech.[127] At the point where the curve breaks and becomes horizontal, the failure mode changed abruptly from shear to tensile failure. At this point the shear load of the matrix or interface is equal to the tensile load of the fiber. Beyond this point the fiber fracture occurs rather than pull-out and the critical L/D ratio is approximately six at room temperature. In actual practice L/D ratios must equal or exceed 100 to achieve theoretically expected strengthening values, on a practical base. In Figure 74 the results are shown of Kelly and Tyson[133] for a tungsten fiber-copper composite as a function of volume loading and L/D ratio at 1212°F (600°C). Experimental points were obtained up to 50 vol % and extrapolated to calculated higher volume fractions. At this relatively low use-temperature the L/D ratio requirement is thus at least 100.

As the shear strength of the matrix decreases at elevated temperatures the need to provide adequate shear strength through longer and longer discontinuous filaments in a tungsten-reinforced copper composite becomes rather drastic. As seen in Figure 75,[127] the required L/D ratio to prevent fiber pull-out increases rapidly above 1400°F (760°C). Considering that major interest in higher temperature systems would include a minimum of 2000°F (1093°C), discontinuous filament reinforcement becomes unattractive, at least in structural materials.

The physical difficulties in misalignment, to return to our earlier argument, make the problem even more serious. Maintaining accurate alignment

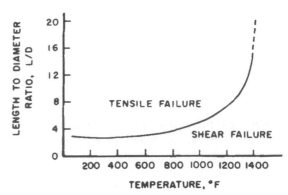

FIGURE 75. Experimental length to diameter ratios for tungsten reinforced copper as a function of temperature.

of short fibers, parallel to the reinforcing axis, is difficult indeed. The results of misalignment at 1500°F (815°C) are shown as a function of the angle of reinforcement to the tensile axis in Figure 76.[132] Similar results will be shown later for misalignment in continuous fiber composites of boron in metals such as aluminum. The difficulty here is in devising experiments and processing methods which control this misalignment. When the misalignment does occur, experimental points agree with the theoretical expectation that at a misalignment angle of 3° at 1500°F (816°C), very little reinforcement is obtained as the vol % of tungsten is increased in a tungsten-copper composite. These results are indicated in Figure 77[132] for a composite with L/D of 100, comparing the axially aligned and 3° misaligned fibers. When the fibers were parallel to the axis the failures were all tensile. With the 3° misalignment, only shear failure was observed. At intermediate

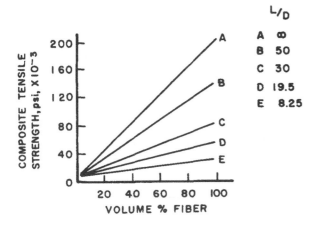

FIGURE 74. Ultimate tensile strength of tungsten reinforced copper composites at 1112°F (600°C), using 8-mil tungsten.

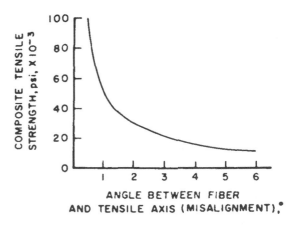

FIGURE 76. Strength as a function of critical angle for tungsten reinforced copper at 1500°F (815°C).

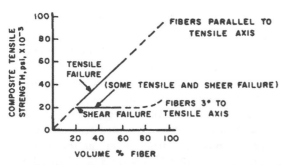

FIGURE 77. Change in strength with volume and orientation of tungsten fibers in copper at 1500°F (815°C). Length to diameter ratio of 100.

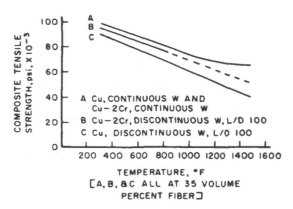

FIGURE 78. Comparison of continuous and discontinuous tungsten reinforced copper composite strength.

values for alignment and a volume loading of 40% fiber, some mixed failures were observed. Given the limitations of these problems with discontinuous fibers and the aggravated handling and fabricating problems from their use, the balance of interest has been moving to continuous filaments for reinforced refractory metals and superalloys.

Mention should be made of the earliest report of a "synergistic effect" in model system studies, wherein the measured tensile strengths slightly exceeded calculated values based on the rule of mixtures. Piehler reported these results[134] for steel fibers in a silver matrix. Alloying reactions in the matrix and fibers and work-hardening as a result of metallurgical processing during fabrication were not considered. There is also considerable variation in experimental results for the strength of fibers and continuous filaments. Therefore, the conclusions of "unusual" strengths do not necessarily have much significance except to indicate satisfactory consolidation techniques. Metallurgical interactions and processing problems such as fiber breaks ordinarily result in lower strengths than predicted.

Copper-Chromium Alloying

Alloying has been employed to improve the properties of both the matrix and the reinforcing wires. The latter will be discussed subsequently. Petrasek[59] alloyed copper with 2% chromium to increase the matrix strength. This increased the reaction zone at the interface with the tungsten wire during fabrication. The increased reaction of the tungsten wire with the copper-chromium alloy decreased the fiber strength to approximately the same degree as the alloy increased the strength for continuous fiber composites. This is indicated in Figure 78, where the 2 matrices give virtually the

same strengths (curve A) with 35 vol % continuous filament and tungsten reinforcement. On the other hand, the increase in matrix strength more than offsets the loss of fiber strength from reactivity upon alloying for the discontinuous tungsten reinforcement. One should note, however, that over the entire temperature range investigated both the Cu and Cu-2Cr matrices gave greater strengths with continuous filament reinforcement. Actually, Petrasek[59] shows a small difference for the two composites represented by curve A, but it is not significant and the results are virtually the same within a reasonable experimental error.

Other Model Systems

Without attempting to be in any way inclusive, some mention should be made of some of the other recent work on refractory metal systems. Karpinos et al.[135] have studied the tungsten-mesh wire reinforcement of nickel. Nickel foil 2 mil thick was hot-pressed with the tungsten grid to prepare composites containing 24 or 38 vol % tungsten. The pressing time at 1200°C of 8 to 10 min seems rather short and the resultant materials have porosities of 2.7 to 3.9%. Marked reinforcement was reported from 20 to 1100°C compared to pure nickel. Insufficient detail has been reported in this work to make much comparison with other efforts, but the network reinforcement would give the advantage of two-axial properties rather than uniaxial reinforcement. Without more information on the compatibility problem and overcoming of porosity during fabrication processes, this work is not very helpful.

Ahmad and Barranco[136] have used continuous tantalum filaments to reinforce copper. This is a ductile filament-ductile matrix system prepared by liquid infiltration. Composites were prepared with up to 38 vol % of 2 mil and 3 mil tantalum wire infiltrated in a 10^{-3} mm Hg vacuum at 1150°C. The composites fracture at a much higher strain than the fracture strain of the filament, showing elongations of 25 to 50%. This corresponds to up to a 30% reduction in diameter of the 3 mil tantalum filament. The authors point out that predicting the strengths of the resultant composites requires considering work-hardening of both the matrix and the filament. Nonetheless, they have attributed significance to the increased axial tensile strengths achieved above a rule of mixtures prediction, without considering these factors. Further, the data they have obtained on the tantalum filaments show considerable variance for various specimens experimentally. A 3 mil filament with an as-received yield strength of 93,000 psi had a strength of 85,000 psi after recovery from a cast composite. A 10 mil filament had a value of 63,000 psi after annealing for the yield strength and increased to 68,000 psi after recovery from a cast composite. This type of variance makes firm calculations of composite strengths difficult, since there are no good "average" values. The tensile strengths of tantalum filaments were found to drop rapidly with increasing size (140,000 to 180,000 psi for 2 mil Ta, for example) to 5 mils diameter, and then level off. Only a small amount of interaction was observed from this liquid infiltration consolidating process.

Another refractory metal wire system recently studied is the reinforcement of columbium (niobium) with TZM (molybdenum alloy) wire to increase the creep resistance of columbium (niobium) alloys. Reece has reported[137] on the explosive bonding of a typical niobium alloy (C129Y), containing substantial amounts of tungsten and hafnium and lesser amounts of tantalum, zirconium, and yttrium. The 10 mil wire had a tensile strength of 252,000 psi and an elastic modulus of 46×10^{-6} psi compared to a strength of 94,000 psi and modulus of 16.4×10^{-6} psi for the 13 mil sheet C129Y alloy. Some improvement in tensile strength and modulus and a sharp drop in elongation were achieved by the explosive bonding process to form a Mo-reinforced Nb composite. Such a composite with improved creep properties is of interest in possible space-shuttle vehicles. One major disadvantage besides the high

density of the resultant composite is the poor oxidation resistance of the components. The TZM wire is even poorer in this regard than the niobium alloy matrix. There is also the question of reactivity between the phases during any long-term elevated temperature exposure.

In a nickel matrix the severe reaction of TZM wire with nickel matrices has been prevented using a vapor deposited barrier layer on the wire of tungsten or alumina.[138] Coatings of 25 to 50 μm thickness eliminated all visible reaction after several hundred hours at 1000°C. Pure Mo wire and another high strength alloy wire of TZC were also tried with similar results. Niobium alloys containing high amounts of tungsten reacted rapidly with nickel base alloys. The reaction was prevented by preoxidation to form an adherent oxide scale of 4 μm thickness on the niobium alloy wires. The effects of these barrier layer coatings on bonding and mechanical properties were not examined. The interaction of tungsten wire with a nickel base alloy (Nimocast 713C) is shown in Figure 79 after 600 hr at 1100°C. The interaction shown by the quantitative microprobe analysis is definite and significant. The results are confirmed with optical microscopy. As previously determined by Dean,[139] the interaction layer at 1000°C after 500 hr is approximately 2 μm, which is about the

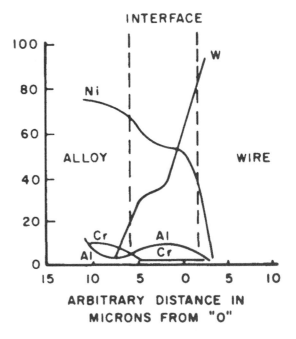

FIGURE 79.   Analysis of reaction between tungsten and nimocast 713C after 600 hr at 1100°C.

same as the thickness in the as-fabricated condition for the composites. The materials were prepared by vacuum casting. Tungsten wire containing 5% rhenium was substituted for tungsten and gave similar results. The first section of the paper by Restall et al.[138] fails to mention any of the previous compatibility work on silicon carbide reinforced nickel, thus repeating it with far fewer results and inadequate understanding of the basic phase problems involved. The results with silicon nitride reinforcement were equally disappointing and the discussion again fails to draw needed conclusions in terms of phase equilibria.

Mention should be made here, perhaps, of attempts to make refractory metal oxide reinforced refractory metal composites. Tungsten metal has been successfully reinforced with both UO_2 and ZrO_2.[140] The refractory oxide powders were extruded in tungsten at temperatures of 2800 to 4600°F (1540 to 2550°C) with the best fibering of the oxide achieved with UO_2 at 3600°F (1980°C). The method makes it difficult to control fiber length (which is discontinuous) and spacing. The resultant fibers are of relatively low strength and modulus, and at least for 20 vol % oxide reinforcement do not compare favorably with high strength tungsten alloys commercially available. Some reinforcement over the unalloyed tungsten was noted, but evidence of brittle behavior with multiple cracking in the matrix at failure was observed at temperatures up to 3600°F (1980°C). Other work reported on niobium reinforced with ZrO_2 fibers using the extrusion process has given essentially the same results.[127] At temperatures from 400 to 2800°F (205 to 1538°C) the ZrO_2 strengthened composites had tensile strengths exceeding those of unalloyed niobium but substantially lower than those for commercially available high strength niobium alloys. For the present, at least, this does not appear to be a promising approach to strengthening refractory metals.

Refractory Metal Wires

After its use in a number of model system studies such as those on the copper-tungsten system, it would be expected that tungsten wire should be considered as a reinforcement in more detail for possible applications. The stress-rupture properties for practical high temperature composites were considered of key importance. McDanels and Signorelli[141] studied the stress-rupture properties of 5 mil diameter, type 218CS lamp grade tungsten wire. At test temperatures from 1200 to 2500°F (649 to 1371°C) for test times up to 200 hr, the stress-rupture properties were superior to those of other forms of tungsten, other refractory metals, and superalloys. Recrystallization at temperatures above 1800°F (982°C) was observed with a subsequent loss in ductility and coarsening of grain size. The calculated stress to give rupture times of 1, 10, and 100 hr for 5 mil drawn tungsten wire is given in Table 9.[141] The drop in stress-rupture life is fairly slow and constant until temperatures above 1800°F where the slope of the stress-temperature curve for a given lifetime increases with the onset of recrystallization. Nevertheless, it is apparent from Figure 80 that the 5 mil wire is far superior to other forms of tungsten for reinforcement. Only at very high temperatures above 3500°F (1927°C) does the form of tungsten become unimportant.

In Figure 81 the stress required for rupture in 100 hr as a function of temperature has been compared for the 5 mil tungsten wire and several refractory metal alloys and superalloys.[141] While the data for the other materials are not for any optimum wire form, the use of tungsten alloys rather than tungsten lamp wire could be expected to yield the type of increase in properties as the difference in TZM over unalloyed molybdenum. There is a serious problem with alloying as already cited for TZM, in that the results may be deleterious in terms of reactions between filament and matrix at high temperatures. (In other cases, particularly with lower temperature systems like titanium, alloying may be used to control these interactions.) So, with some limitations, at least Figure 81 shows that the tungsten wire compares

TABLE 9

Rupture Life for 5-mil Tungsten Wire

Test temperature		Stress for rupture, ksi		
°F	°C	1 hr	10 hrs	100 hrs
1200	649	199.1	175.5	154.8
1500	816	145.6	130.4	116.7
1800	982	116.0	104.5	94.2
2000	1093	88.1	73.3	61.0
2300	1260	62.8	49.0	38.3
2500	1371	49.1	35.9	26.2

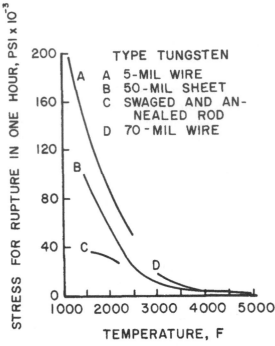

FIGURE 80. Stress required for rupture in 1 hr at various temperatures for different forms of tungsten.

favorably with other types of potential reinforcements.

In further work on the tungsten reinforced copper model system, McDanels, Signorelli, and Weeton[142] found that the stress-rupture values for the 5 mil tungsten wire were affected by the fabrication treatment conditions. No significant interaction occurred between the copper and

tungsten, but annealing the wire at the infiltration condition resulted in a change in stress-rupture life of the wire as shown in Figure 82. While the model systems studies had shown that creep and stress rupture properties up to 1500°F (815°C) compared favorably with those of existing superalloys, the problem of reactivity in selection of the fiber reinforcement for practical systems posed a serious problem. The most direct approach to minimizing the loss of strength of the fiber in alloying reactions is to utilize larger diameter wire so that the deleterious interaction involves less of the filament volume. The rate of penetration of the matrix at the filament surface was found to be relatively constant regardless of the fiber diameter,[143] and the fraction of fiber area reacted with time less with increasing fiber diameter. The limitation in this approach is seen in Figure 80. The smaller diameter wire is generally stronger than the larger diameter wire. This trade-off in diameter and strength has resulted in increased experimentation with larger wires of 15 to 20 mils in diameter.[143] Petrasek, Signorelli, and Weeton[143] tried several refractory wire compositions in their early work on reinforcing superalloys. These included TZM (Mo-0.5Ti-0.08Zr-0.015C), Tungsten-1 ThO_2, Tungsten-3Re, and commercial tungsten. Several nickel alloys were studied to determine an optimized composition where the reaction rate was minimized. Small amounts of Al and Ti were beneficial to obtaining lower reaction rates, while the amount of W in the Ni alloy was not critical. The most compatible alloy found was Ni-2Al-15Cr-2Ti-25W. Good results were also obtained with an alloy containing Al and Ti, but only 4W. Poor results were obtained with alloys having no Al and Ti additions. The depth of the penetration reaction between the matrix and the filament for the most compatible nickel alloy and 8 and 15 mil tungsten reinforce-

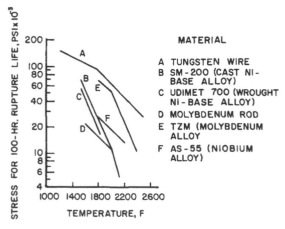

FIGURE 81. Stress required for rupture in 100 hr at various temperatures for different refractory metals and alloys.

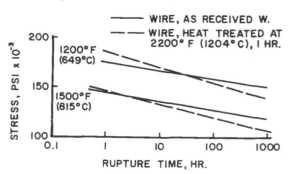

FIGURE 82. Stress-rupture life of tungsten wire.

ment, as a function of stress-rupture, is shown in Figure 83. Such studies can be used to decide on the amount of reaction that can occur and still have the composite retain satisfactory properties for a given application. A simple graphical method has been employed to illustrate schematically the change in optimum wire diameter with allowable penetration limits (and composite strength), as shown in Figure 84.[128,143] In compatibility terms the W and W-Re wires were more attractive than the W-ThO$_2$ and TZM wires. This calculated optimum wire size as that wire diameter yielding the highest fiber stress contribution to the composite at a specific reaction depth has been partly verified by 100 hr exposure at 2000°F (1093°C) of W reinforced Ni alloys. An optimum diameter of 8 to 15 mils would seem indicated as was given in the first work.[143] Later this was

placed at 15 mils,[128] but given at 20 mils for the W-ThO$_2$ wire. Since alloying has such a marked effect on the depth of penetration, and considering that the actual effect of alloying on the stress-rupture properties of the wire is not really known, this approach to optimize wire diameter must be considered only a qualitative tool. It is nonetheless very useful to the practitioner.

The larger diameter W-ThO$_2$ and W lamp wire,[23] and a combination W-ThO$_2$-Re wire have been investigated in more detail with the attempts to develop practical wire-reinforced nickel base superalloys. The tensile strengths of W-2ThO$_2$-5Re, W-2ThO$_2$, 218 CS W (lamp wire), and W-1ThO$_2$ at room temperature to 2200°F (1204°C) are given in Table 10. The stress-rupture properties of these same wires are given in Table 11 at the temperatures of 2000°F (1093°C) and 2200°F (1204°C). Data for tensile tests were obtained at 10^{-5} mm Hg using a constant-strain-rate, screw-driven crosshead, universal testing machine at a crosshead speed of 0.1 in./min (0.25 cm/min). Stress rupture tests were obtained in a vacuum chamber at 5 x 10^{-5} to 1 x 10^{-6} mm Hg with the wire strung through a tantalum-wound resistance furnace, around a pulley, and attached to a weight pan. The temperature was stabilized before applying weights. The tensile strength of the W2-ThO$_2$ wire is somewhat surprisingly higher at the 15 mil diameter than at the 10 mil diameter. The wire containing Re has excellent strength but is not as attractive in stress-rupture life. The 218 CS lamp wire and the W-1ThO$_2$ wire are low both in tensile and stress-rupture properties. Some of the data obtained on the W-1ThO$_2$ wire in stress-rupture at 200°F are given in the original reference[23] with considerable variation and not included here. For example, at 55,000 psi the stress-rupture life was given in two tests as 33.9 and 12.2, or almost a 300% difference from the lower value. It is clear, however, that the 2% thoria addition is extremely beneficial to the high temperature properties.

Weeton and Signorelli[144] have reported on continuing efforts to obtain a wider selection of refractory metals in wire form for use in reinforcing superalloys. Investigations are being conducted on chemical vapor deposition, ion plating, and other methods of coating tungsten and developing other reinforcing materials. A new niobium alloy based wire has been developed, but

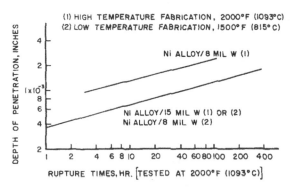

FIGURE 83. Effect of fabrication process and wire diameter on depth of penetration in stress-rupture test at 2000°F (1093°C).

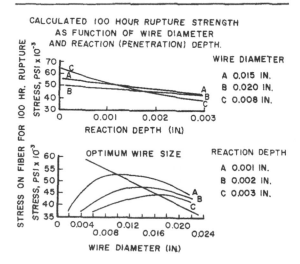

FIGURE 84. Stress-rupture calculations for tungsten wire at 2000°F (1093°C).

TABLE 10

Tensile Strength of Refractory Wires

Wire	Diameter, in.	Strength, ksi		
		70°F (21°C)	2000°F (1093°C)	2200°F (1204°C)
W-2ThO$_2$-5Re	0.020	310	185	147
W-2ThO$_2$	0.010	399	147	132
	0.015	384	173	150
218CSW	0.015	346	111	94
W-1ThO$_2$	0.020	335	116	107

TABLE 11

Stress-Rupture Properties of Refractory Wires

Wire	Stress, ksi	Rupture life, hr	Temperature
W-2ThO$_2$-Re	100	18.4	2000°F (1093°C)
(0.020 in.)	90	41.2	2000°F (1093°C)
	85	85.3	2000°F (1093°C)
	80	1443	2000°F (1093°C)
	80	4.9	2200°F (1204°C)
	60	17.1	2200°F (1204°C)
	50	55.3	2200°F (1204°C)
	40	174.1	2200°F (1204°C)
W-2ThO$_2$	120	14	2000°F
(0.010 in.)	110	24	2000°F
	105	61	2000°F
	100	328	2000°F
	90	13.4	2200°F
	80	34.1	2200°F
	75	66.4	2200°F
	70	118.7	2200°F
W-2ThO$_2$	100	6.5	2000°F
(0.015 in.)	97	43.3	2000°F
	95	228.1	2000°F
	93	218.5	2000°F
	80	17.8	2200°F
	75	26.1	2200°F
	70	105.1	2200°F
	60	146	2200°F
W-218CS	80	4.0	2000°F
(0.015 in.)	70	9.5	2000°F
	60	44.2	2000°F
	55	86.8	2000°F
	40	105.9	2200°F
W-1ThO$_2$	90	5.1	2000°F
(0.015 in.)	85	7.7	2000°F
	83	23.4	2000°F
	80	133.5	2000°F
	60	20.4	2200°F
	50	79.3	2200°F
	45	279.1	2200°F

with properties just slightly better than W-2ThO$_2$. More promising is the recently reported[144] W-Hf-C wire with substantially higher strength at elevated temperatures.

Superalloy Composites — Practical Systems

For reinforcing phases of high density such as refractory metal wires represent, the major interest is expected to be in high temperature reinforcement. Therefore, the reinforcement of superalloys for use at higher fractions of their melting points has been the focus of much of the more pragmatic investigations. The potential of this type of reinforcement is indicated in the schematic Figure 85. The shaded area represents the range of tensile strengths at various temperatures for the metal composites. The lower curve is for superalloys (several years ago) and then an arbitrary 25% increase in strength has been indicated with another curve. Other properties of critical importance such as stress-rupture life and creep resistance have been found related to tensile strength for previous work, so the curve could use several different properties vs. temperature. It is seen that at precisely the higher temperature range from 2000°F (1093°C) to 2400°F (1315°C) the composites are most attractive. It then becomes a matter of reactivity as a limiting factor for useful temperature and useful life. The reinforcement must have substantially higher strengths than the superalloys they reinforce, which is particularly true at the very high temperatures (above 1600°F (871°C)) where the matrix strength falls rapidly compared to refractory wires such as tungsten. As

we have seen, alloy wires of tungsten are even more promising, and diffusion barriers and judicious alloying may offer some improvement in the reactivity picture.

In Figure 85 the fiber reinforced composites have been shown in a range consistent with the use of wire reinforcements with considerable variation in initial strengths and the use of low (10 to 20) to high (70) vol % reinforcement phase. The development of new, stronger fibers — which is certainly a good possibility considering how little has been done — would push the upper values still higher.

Using tungsten and tungsten alloy wire reinforcement of nickel base alloys, Petrasek and Signorelli[23,129] have demonstrated that practical composites of this type can be fabricated. Much of the earlier basis for this effort has been reviewed by Weeton and Signorelli.[127,128] Petrasek and Signorelli[129,23] prepared the composites by a slip casting technique with the nickel powder slip and aligned tungsten fibers placed in a nickel tube on a vibrating table.

The composites were dried in air for 24 hr at 140°F (60°C) and sintered at 1500°F (815°C) for 1 hr in dry hydrogen. The specimens were then isostatically hot-pressed in Inconel tubes at 20,000 psi with helium at 1500°F (815°C) for 1 hr and then 1 hr at 2000°F (1093°C). Composite specimens containing 25 to 70 vol % fiber were fabricated in this manner which were virtually at theoretical density (> 99%).

The potential of W-2ThO$_2$, 218 CS W, and W-1ThO$_2$ to reinforce a nickel base superalloy for turbine applications was evaluated based on stress-rupture strength, oxidation, and impact resistance. The W-2ThO$_2$-5Re wire[23] was not investigated in as much detail because the W-2ThO$_2$ wire is much stronger. The alloy used was the previously developed compatible Ni-25W-15Cr-2Al-2Ti composition. The results indicated that refractory metal alloy fiber-superalloy composites have potential for turbine bucket use based on the properties measured. Composites were produced having stress-rupture properties superior to conventional cast superalloys at use temperatures of 2000°F (1093°C) and 2200°F (1204°C). The 100 hr and 1,000 hr stress-rupture strengths of the composite at 2000°F (1093°C) were 49,000 and 37,000 psi, respectively. This compares with 11,500 and 6,000 psi for cast nickel base alloys under the same test conditions. A few thousandths of an inch of matrix material was found to be

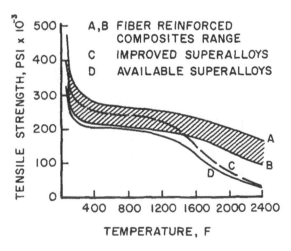

FIGURE 85. Potential strengths of refractory composites compared to superalloys.

sufficient to protect the fibers from oxidation at 2000°F for times up to 300 hr. At 300°F (149°C) and above the impact resistance compares favorably with that of the superalloys.

The impact studies were made on composite specimens having fiber contents of 50 to 60%, but only with the 218CS and W-1ThO$_2$ compositions in the nickel alloy. A Charpy impact tester for tests at 2000°F (1093°C) and a low capacity Izod impact tester for tests at room temperature to 300°F (149°C) were used.[129] The Charpy specimens were machined to the ASTM Specification for Notched Bar Impact Testing of Metallic Materials (E23-64), or 0.394 x 0.394 x 2.165 in. Izod specimens were one half the ASTM specification size.

Above 300°F (149°C) the impact resistance of these composite specimens compared favorably with that of the superalloys.[129] The low impact values observed for the composites tested below 300°F were attributed to the brittleness of the tungsten wires below their transition temperature as indicated in previous work.[59] Improvement in room temperature impact strength should be possible, although impact values obtained are considered sufficient by Petrasek and Signorelli to warrant consideration for use of this type composite in turbojet buckets.

Oxidation studies were conducted on the composites which were completely encased in an Inconel cladding 0.014 in. thick which was ground down to a 0.006 in. (0.015 in.) cladding on the cylindrical specimens. Two other specimen types were prepared: (1) both ends were machined away to expose the cross section of the wire reinforcement to the atmosphere; (2) both the ends and the sides were machined away to expose both cross section and longitudinal wire sections to the atmosphere.

Static air oxidation tests were conducted to 2000°F (1093°C) at times up to 300 hr or more. The completely clad materials showed oxidation of the Inconel with formation of a coherent oxide scale and complete protection of the fibers under these conditions. The tungsten wires exposed in either direction oxidized rapidly, but in the transverse direction, any wires not exposed were protected by the matrix. Since the best superalloys may also require additional oxidation protection for the higher temperature regime, coatings being made for this use are considered promising for possible fiber composites cladding and should be

more effective than Inconel used in these experiments.[23]

Stress-rupture properties measured for these refractory wire reinforced nickel alloy composites were superior to existing materials. The best values were obtained with the large diameter, 0.015 in. (0.038 cm), W-2ThO$_2$ wire composites. The results for 100 and 1,000 hr stress-rupture strengths are shown in bar graph form in Figure 86.[23] The stress-rupture strengths for 70 vol % W-1ThO$_2$ and W-2ThO$_2$ reinforced nickel alloys have been compared with the rupture strengths for conventional cast nickel alloys at 2000°F (1093°C). The 100 and 1,000 hr strengths are significantly higher for the composite structures with the W-2ThO$_2$ much better than the W-1ThO$_2$ reinforcement. The W-2ThO$_2$ reinforced nickel alloy composite had a 100 hr rupture strength 4 times that of the strongest conventional cast superalloys at 2000°F and a 1,000 hr rupture strength 6 times that of the superalloys.[23]

The density of the composite made with tungsten wire is considerably greater than that of cast nickel alloys. Tungsten itself is over twice as dense as most nickel alloys, and a composite containing a high volume fraction of tungsten reinforcement is heavy. In Figure 87,[23] the density has been accounted for in comparing specific rupture strengths of the composite structures and conventional cast nickel alloys. While the margin of property superiority is less, the W-2ThO$_2$/Ni has a 100 hr rupture strength at

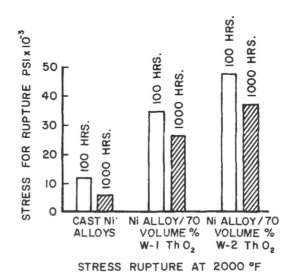

FIGURE 86. Comparison of rupture strengths of cast nickel alloys and refractory composites.

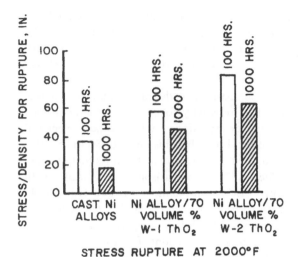

FIGURE 87. Comparison of specific rupture strengths of cast nickel alloys and refractory composites.

2000°F (1093°C) of 83,000 in., which is over twice that of the superalloys. At 1,000 hr the rupture strength of W-2ThO$_2$/Ni is 63,000 in. at 2000°F, which is over 3 times that of the superalloys. This comparison is indicative of the potential for use of refractory wire reinforced composites for higher temperature applications such as turbine blades. Standard nickel alloys in this application have a stress-density value of about 60,000 in. for a 1,000 hr rupture life at 1800°F (982°C). The refractory wire composite exceeds this value slightly at 2000°F (1093°C), so there is a use temperature gain of 200°F (93°C) in favor of the composite over conventional superalloys.[23] Considering that only matrix compatibility served for matrix selection in brief study, improvements in obtaining a stronger (and still compatible) matrix can be expected. The behavior in thermal cycling and compatibility over long periods of time at temperature remains to be investigated.

Another recent investigation of refractory wire reinforcement is that of Klein, Domes, and Metcalfe[145] on tungsten fiber reinforcement of niobium alloys. In this study weak oxidation-resistant matrices of niobium alloys have been reinforced with W-3Re wire filaments of 0.010-in. diameter. A model alloy system to determine feasibility was Nb-40Ti-9Cr-4Al. A number of other alloy substitutions were also employed with 10W, additions and changes in Cr content from 4 to 7%. Reasonable properties were obtained above

2000°F (1093°C). The longitudinal strength for 24 vol % tungsten was 170 ksi at room temperature and 50 ksi at 2200°F (1204°C). The transverse strength decreased from 80 ksi at room temperature to 6 ksi at 2200°F. Stress-rupture life was 250 hr for stresses up to 25 ksi for a Nb alloy/24W composite in air at 2200°F. Kirkendall porosity in the W-based fiber, low transverse strengths, matrix embrittlement, and substantial matrix and filament oxidation at even rather low temperatures were identified as some of the major problems. They are currently under study. Whether a sufficiently oxidation-resistant matrix or suitable coating can be developed remains to be seen.

A number of other attempts at refractory reinforcement utilizing tungsten in superalloy matrices have been studied and have been reviewed elsewhere.[128]

Brennan[146] selected one particular composition of a niobium alloy from the work of Klein, Domes, and Metcalfe,[145] and studied the use of plasma spray tape fabrication methods. A Nb-40Ti-10Cr-5Al alloy was reinforced with 218 CS-5 mil tungsten wire and fabricated into a plasma tape, without serious degradation of the composite. The plasma tapes were diffusion-bonded into larger specimens by hot pressing and preliminary strength data were obtained. Reported tensile strengths for a 50 vol % tungsten composite were 142,000 psi at room temperature and 60,000 psi at 1800°F (982°C). These values are definitely inferior to the results for the tungsten reinforced Ni alloys and for Klein et al. on Nb alloys. This is due partly to the use of the smaller tungsten filament, perhaps, with a loss of strength during fabrication interactions.

Reinforced Electrical Conductors

One area where the tungsten reinforced copper system has practical interest is as a high strength electrical conductor. The electrical properties of copper composites with 14 to 75 vol % tungsten were studied by McDanels.[147]

Since the components of the composite are mutually insoluble, McDanels expected the electrical properties to follow those of binary alloys such as Pb-Sn and Zn-Cd, and have a linear variation of conductivity with composition. The experimental conductivity of the tungsten-copper composites lay on a line between the conductivities of pure copper and tungsten. Since the strength of tungsten wire is approximately 11 times and the resistivity is approximately 3 times that of copper, a compromise between conductivity and strength can be made. High strength electrical conductors are promising based on this approach.

Other investigations have also indicated the practicality of fiber reinforcement of copper to develop high strength electrical conductors,[148] since higher conductivity materials are generally weak, and stronger materials are poorer conductors. (This is with the notable exception of certain transition metal compounds such as titanium diboride (TiB_2) which are excellent conductors and very strong, but, unfortunately, quite brittle.) Composites have been made from copper and silver matrices reinforced with fibrous felts of tunsten and other metals. A linear increase in resistivity with fiber content to a maximum of 40 vol % fiber was reported.[149] Steel-reinforced (ACSR) aluminum conductors have also been fabricated which have strands of low-density, high conductivity EC aluminum reinforced by strands of relatively low-conductivity steel of a similar diameter.[150] This is a composite cable because it has two distinct components; they are not bonded but rely on the friction of stranding to gain reinforcement of the aluminum with the steel strands. Such reinforcement increases the tensile strength to resistivity ratio substantially above that of EC aluminum. When the density is taken into account the specific strength to resistivity ratio is higher for composite containing six strands of aluminum to one strand of steel, but about the same as EC aluminum when a ratio of three strands of aluminum to one strand of steel was used.[150]

The comparison of the ratio of ultimate tensile strength to resistivity for tungsten fiber reinforced copper composites with other electrical conductors shows that this ratio for 50 to 75 vol % tungsten wire is three to seven times that of other conductors.[147] Comparisons were made with silver, EC aluminum, oxygen-free high-conductivity copper, and steel-reinforced aluminum cable. When conpared on a specific strength basis to account for the high density tungsten, the maximum ratio for the ultimate strength/density/resistivity occurs at 50 to 60 vol % tungsten wire and is 30% greater than for aluminum, 100% greater than for copper, and 15% greater than steel-reinforced aluminum cable. The strength/resistivity ratio of tungsten copper composites increases from 10 vol % to a maximum of 70 vol % tungsten wire and then falls off again. The specific strength increases from 10 vol % to the 50 to 60 vol % tungsten wire range and then falls off. In applications where external loads such as wind, ice, and foreign objects are concerned, the advantage of the tungsten fiber reinforced copper composites over other conductors is clearly apparent.[147] A sixty vol % reinforcement of tungsten wire would be best based on the reported data.

Chapter 6

WHISKER REINFORCED METALS

Of all reinforcing fibers applied or considered for application in metal matrix composites, short single crystal filaments called "whiskers" are by far the strongest. The whisker sizes generally range from 1 to 10μ cross section and from a few mm to several cm in length. With strengths up to several million psi, they represent the nearest approach achieved in the laboratory toward realization of the theoretical strength of matter, as calculated from the strength of the interatomic bonds.

Interest in their practical utilization in metals can be traced to the work of Brenner, who grew relatively large α-alumina (sapphire)* whiskers and determined their room and elevated temperature properties.[151] Brenner suggested their use for reinforcing of a metal matrix.[152] He calculated the fiber-matrix shear strength required to achieve fracture of a composite containing aligned whiskers through massive whisker fracture rather than whisker pull-out. Hahn and Kershaw[153] point out that Brenner's assumptions and results were, by present standards, somewhat over-simplified; but Brenner did use the important concept of "critical aspect ratio" (the ratio of length to diameter of a short fiber) which for a given value of fracture strength and metal-fiber shear strength made the strength of a composite dependent on whisker-strength rather than whisker-matrix shear strength. The theory of whisker reinforcement was subsequently refined[154] to consider the elastic and plastic properties of both the whisker and the matrix in the calculation of critical aspect ratios. This theory predicted a significant and often drastic increase in whisker length requirement with increasing temperature. Fortunately, some whisker growth processes yield whiskers with aspect ratios of 10,000:1 or more.

Early experimental work was reported by Sutton,[155] who demonstrated successful reinforcement of a silver matrix with alumina whiskers to approximately 98% of the melting point of silver. His work gave impetus to further investigations to demonstrate such reinforcement in more practically useful metals and alloys. Simultaneously, processes were developed for more economical sapphire whisker growth and for the growth of silicon nitride and silicon carbide whiskers (both the low temperature alpha and high temperature beta forms).

Sutton[99] was able to demonstrate the experimental development of alumina-whisker reinforced nickel for possible use at temperatures exceeding 1832°F (1000°C) for extended periods of time. Composite efficiency in utilizing the high strength of the reinforcing "whiskers" was not very good, however.

The difficulties in achieving a bond between alumina whiskers and the metal matrix made whisker-coating necessary prior to consolidation. The same problem occurred with aluminum as with nickel, but to a lesser extent. Mehan[156,157] has studied the alumina reinforcement of aluminum and obtained encouraging high temperature properties. Composite strengths, however, were again much lower than expected.

Advantages of whisker composites include potential consolidation by modified powder metallurgical or casting processes. Orientation of the whiskers can be uniaxially parallel to a single axis, randomly distributed in a plane, biaxial, or in varying longitudinal to transverse ratios. The composites can be hot worked by rolling, forging, or extrusion. Machining is accomplished by conventional means utilizing carbide tooling. Drilling, grinding, and many joining techniques do not degrade mechanical properties. Whisker-composite strengths are lower than the best high volume fraction continuous filament composites. However, both types of composites are of comparable strength at volume fractions below approximately 20 vol % reinforcement. In addition, several whisker alloys have been produced with excellent stress-rupture and fatigue properties, which are preserved to well above the upper temperature limit of the best high temperature aluminum and magnesium alloys.[153]

Whiskers

Alpha alumina whiskers are prepared by a vapor

*Single crystal α-Al_2O_3 often referred to in the literature as sapphire whiskers.

phase reaction in which moist hydrogen is passed over molten aluminum. Temperature, time, system geometry, and composition of the atmosphere are used to control growth rate, morphology, and aspect ratio. The strength is approximately 10^6 psi for whiskers having a cross section of approximately 1μ, and their modulus is approximately 70×10^6 psi. The strength drops with increasing cross sectional area at an average rate so that a whisker with a diameter of 10μ would be expected to exhibit an average strength of 800,000 psi. The scatter in observed strengths has been large.[3] Some properties of available whiskers are given in Table 12. Tensile strength is maintained to high temperatures, as indicated in Figure 3. Alumina is inert to many structural metals at relatively high temperatures.

Silicon carbide whiskers are currently available in somewhat greater quantities and are generally cheaper. The alpha form is grown by means of a process wherein organosilane is thermally decomposed in hydrogen at a temperature above $1400°C$.

A low-cost process[153] utilizes solid phase elemental silicon and carbon in an atmosphere of hydrogen and hydrogen chloride at $1200°C$ to $1450°C$. The whiskers grow by a reaction between $SiCl_2$ and CH_4, the two principal volatile species present.

The properties of silicon carbide whiskers are somewhat similar to those of alumina except that their strength is less diameter dependent and their density is lower. Unfortunately, silicon carbide whiskers are reactive in nickel, cobalt, chromium, and other refractory metals and their alloys.

Silicon nitride whiskers are grown by a vapor reaction between silicon and a silicate in nitrogen gas diluted with hydrogen, at a temperature above $1400°C$. Their properties are lower than those of alumina and silicon carbide whiskers, and their compatibility with most matrices is poor.

The high cost of whiskers as seen in Table 12 mitigates against large-scale use. The projected costs, however, would make whisker reinforcements competitive with continuous filament reinforcements and considerably improve interest in their development.

As-grown whiskers require further processing prior to their incorporation in metal matrices. Such processing must include:[153]

a. Removal of debris, including globular aggregates of the whisker material.

b. Combing out of short or excessively tangled whiskers.

c. Separation or classification by diameter and/or aspect ratio ranges.

The techniques utilized in accomplishing the above vary considerably in accordance with whisker species and also vary from one laboratory to another. The whiskers are usually handled in a liquid medium such as water or an organic solvent, or by air elutriation. The air elutriation process applied to alumina reinforced aluminum was recently described by Mehan.[158]

A process suitable for beneficiating long, as-grown β-SiC wool has been developed.[159] The wool mats were infiltrated with camphene, a low melting hydrocarbon. When cooled to about $5°C$,

TABLE 12

Properties of Commercially Available Whiskers As Furnished by Vendors

Whisker	Vendor	Diameter (microns)	Density (lbs/in.^{3})	Tensile Strength (10^3 psi)	Elastic Modulus (10^6 psi)	Cost ($/lb) Current	Projected
Al_2O_3	A	1-10	0.143	600-3500	80-150	2700-13500	
Al_2O_3	B	0.5-2.0	0.143	1000-3000	60-70	9000	
Al_2O_3	C	0.5-2.0	0.143	1400	64	13000	
α-SiC	A		0.116	350	70	NA	15-50
α-SiC	B	0.5-3.0	0.116	350	70	250	
β-SiC	A	0.1-5.0	0.116	1200-1500	60-70	1800	250
β-SiC	B	1-10	0.116	2000-6000	80-150	9250	250
Si_3N_4	A	1-10	0.115	700-1500	40	1500-3000	

the mass had properties somewhat similar to those of paraffin wax.

The material was then extruded at an extrusion ratio of between 1:80 and 1:200. This process "combed" the whiskers, which were extracted in alcohol and fired to remove all traces of camphene.

The whiskers were then dispersed in water, decanted to remove debris, and liquid screened. The average aspect ratio remained greater than 400.

The development of alumina whisker reinforced composites requires selection of matrices that have the ability to achieve whisker/matrix shear bond strength at elevated temperatures. For silicon carbide and silicon nitride the primary limitation has been chemical incompatibility at elevated temperatures.

Understanding the nature of laboratory problems in working with whiskers is not easy without some experimental work. A few pictures do illustrate these difficulties, however.[25] Whiskers are normally received from commercial sources in the wool mat form, as seen in Figure 88A for a batch of a-SiC whiskers. These clumps of whiskers must be most carefully handled to minimize breakage, damage, and loss while processing to achieve the aforementioned processing goals. The beneficiation of whiskers is as much an art as it is a science, and poses a major

FIGURE 88A. Whisker wool (4x a-SiC).

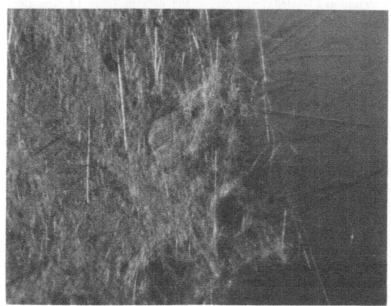

FIGURE 88B. Portion of whisker wool (70x a-SiC).

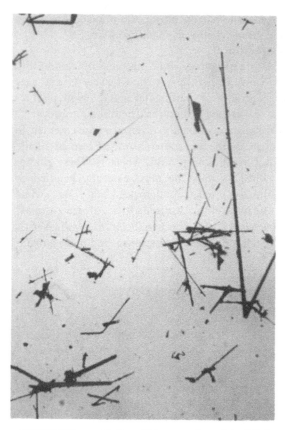

FIGURE 89.   α-Al₂O₃ whiskers on glass plate (70x).

individual whiskers packed together. The range of sizes of whiskers in these clusters is usually quite large. This is observed in Figure 89 for α-Al$_2$O$_3$ *whiskers separated from a whisker mass onto a glass plate.* Since only the larger aspect ratios are desirable for reinforcement, a high percentage of the whisker mass exclusive of general debris must be thrown out. This is a considerable loss in *apparently useful material during beneficiation,* and this significantly increases, in turn, the effective price per pound for whisker reinforcements.

On closer examination many whiskers are found to be more complex than one might expect from apparent single crystals. Electron micrographs of β-SiC whiskers received from one commercial source, for example, have shown a distinct central core under very high magnification. This is shown in Figure 90.

Coating whiskers by vapor deposition methods is commonly employed to improve bonding between matrix and whisker, and sometimes to provide improved compatibility. In Figure 91A a *smooth coating of aluminum on an* α-SiC *whisker* is shown. Figure 91B shows an uneven coating of aluminum on an α-Al$_2$O$_3$ whisker. In each case the aluminized whiskers were employed to improve the consolidation of the whiskers into aluminum alloy matrices. Coatings that are thick or uneven, as for the α-Al$_2$O$_3$ whiskers shown may result in a marked decrease in strength. This is confirmed by the data in Table 13. The aluminum coating on

hindrance to automated methods of consolidation and fabrication of whisker-reinforced materials. The wool shown in Figure 88A is shown in Figure 88B at higher magnification as a tangled cluster of

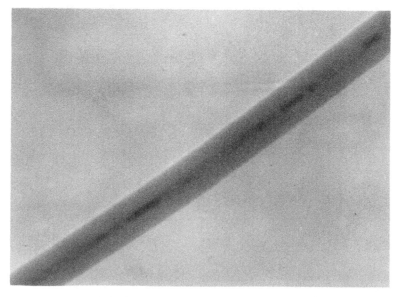

FIGURE 90.   β-SiC electron micrograph (96,000x).

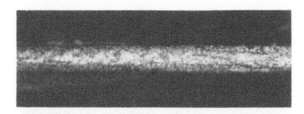

FIGURE 91A. Even coating. Aluminized a-SiC whisker (1070x).

FIGURE 91B. Uneven coating. Aluminized a-Al$_2$O$_3$ whisker (1070x).

these alumina whiskers was thick enough to significantly reduce the elastic modulus values. Considerable scatter in the data was experienced, as is often the case in measurements on whiskers. In every test, though, the trend was identical toward lowering the strength to a substantial degree in the aluminizing process. Where the

coating was very thick, as in Test 2, the strength and modulus were affected much more than where the relative thickness was less, as in Test 3.

Whisker Testing

For many years the AFML conducted considerable characterization of all available filamentary materials. Dr. J. A. Herzog developed many specialized techniques for handling and measuring whisker properties. Many of these techniques were summarized in a series of reports.[160-163] The testing machine which was designed with the necessary small dimensions to accurately study whisker properties is shown in Figure 92. The frame consists of a base plate with two columns connected by a crosshead at the top. In the center, two parallel, round, vertical bars are attached to flat bars at the top and bottom. Between the round bars, a crosshead slides up and down, guided in bushings. The crosshead is moved vertically by a center spindle for adjustment of the load measuring spring to the exact zero position. On the crosshead a support bracket for a 1/1000 micrometer is fastened. The upper flat bar has another bracket for the support of a horizontal plate with a square hole in the center. The specimen block is shown in Figure 92 on the

TABLE 13

Properties of a-Aluminum Whiskers and Aluminized Whiskers

Test	Specimen	Cross sectional area (10^{-6} mm^2)	Elastic modulus (10^6 psi)	Ultimate tensile strength (10^6 psi)
1	coated	48.856	11.67	0.075
	uncoated	12.280	46.43	0.275
2	coated	134.73	22.29	0.145
	uncoated	51.207	58.79	0.673
3	coated	45.997	15.33	0.192
	uncoated	29.759	23.71	0.243
4	coated	15.073	20.02	0.116
	uncoated	9.835	30.67	0.179
5	coated	81.709	74.55	0.260
	uncoated	62.023	98.26	0.332
6	coated	51.661	26.13	0.176
	uncoated	27.446	49.27	0.306

FIGURE 92. Whisker testing machine (scale in inches).

center top clamped with a bracket. Perpendicular to the specimen axis, a light with a front end condenser is mounted (not shown) to illuminate the specimen. A Bausch and Lomb anastigmat lens (shown at top left in Figure 92) with a focusing length of 27 mm is in the light path to produce a focused image of the whisker specimen at approximately 200 in. The image is projected on a small ground glass screen with a precision dial gauge attached to mark gauge mark movements of the whisker specimens.

For load measurement a double coil spring is used with its suspension points 180° out of phase to eliminate moments introduced by the loading mechanisms. These coil springs must be identical and clock hair springs were found to be adequate and economical. The hook on the upper end of the loading spring was made of 0.015 in. diameter piano wire.

The whisker is mounted in a block in wire loops, one suspended on a rigid pin, and the other attached to the hook of the loading spring. For mounting, the wire loops are clamped with Plexiglas® plates which are released when the specimen is ready for testing. The whisker specimen is mounted in the rigid block between the wire loops and attached with cement. Gauge marks are provided by cementing short whisker pieces to the test whisker specimen 1.0 to 1.5 mm apart. Micromanipulators were required for this work under a bench binocular microscope at 20 x. The wire loops were made generally from 0.005 in. diameter piano wire. The entire process requires considerable care, and in practice the measured strengths of whiskers vary considerably. This is shown in the large strength ranges for commercially available whiskers listed in Table 12.

Consolidation

Composites incorporating whiskers (particularly alumina and silicon carbide) and several metals and alloys have been consolidated using laboratory-scale and/or modified industrial techniques. These have included liquid infiltration of aligned whisker bundles, casting, electroforming, high energy rate forming, and a variety of powder metallurgical processes.

Liquid infiltration of hand or slurry process aligned whisker bundles has proved successful on a laboratory scale.[156] Unfortunately, the process applied to alumina whisker systems has no industrial analogy.

Electroforming has been useful in preparing aligned nickel matrix composites, containing as high as 27 vol % alumina whiskers. The high volume fraction was achieved by flow motion. Titanium coated whiskers were successfully electroformed in a nickel bath.[164]

Williford and Snajdr[65] studied the consolidation of whisker composites by a high energy rate process (Dynapac) at room temperature. The systems studied included silicon carbide in titanium and Ti-6Al-4V, as well as alumina in Ti, Ti-6Al-4V, Ni, and Fe-25Cr-5Al. Though fragmentation of whiskers during consolidation occurred in all the systems studied, the alumina-Ni system appeared more feasible than the others for further investigation.

The application of powder-metallurgy techniques to whisker-aluminum or magnesium alloy composites suffers from the lack of purity and availability of sufficiently fine powders to achieve size compatibility with whiskers. The physical process of aligning whiskers is quite difficult. Ideally, the matrix powder should be of smaller particle size than the largest whisker diameter[153] to achieve good orientation by any of the processes generally employed. Another limitation is associated with whisker fragmentation which occurs unless a high degree of pressure and temperature control is exercised. The best results thus far with respect to the whisker fracture problem have been achieved by the liquid-phase hot-pressing technique[165] wherein the matrix and pressing temperature have been selected to yield a small amount of liquid at the highest temperature reached during a carefully controlled hot-pressing cycle.

Liquid-phase hot-pressing of randomly oriented whisker-powder compacts places an upper limit of some 20 vol % on whiskers, beyond which billet densities are insufficient for subsequent hot extrusion. Such extrusion is employed to achieve alignment providing extrusion ratios of the order of 20:1 are used. At lower extrusion ratios poor or no alignment is achieved.

Higher volume fractions — to approximately 50 vol % — have been achieved in liquid-phase hot-pressing of pre-aligned bodies. Two alignment techniques have been investigated. In one, alumina whiskers were coated with nickel by chemical vapor deposition and then liquid blended with the matrix. The suspension was placed in an 8 kG magnetic field, where the liquid was aspirated through a filter upon which the aligned green body was formed. The second technique utilized camphene extrusion of whisker-matrix powder mixtures into square stringers, which were subsequently stacked in a die (longitudinally) and liquid-phase hot-pressed.[153]

While only dilute aluminum and magnesium alloy base whisker composites have been successfully extruded thus far, the results have been encouraging, particularly on the extrusion of aluminum-silicon carbide composites.[159] Such extrusion served to achieve breakdown of the as-cast or hot-pressed billet, to orient whiskers, and/or to provide a desired final shape. No significant reductions have been achieved, however, without an accompanying reduction in whisker aspect ratio, tending to deteriorate elevated temperature properties. A serious limitation here is that the only shapes available are those which can be extruded with large reductions in cross section.

Extrusion of magnetically oriented or camphene formed billets resulted in less whisker fracture than the extrusion of randomly or planarly oriented billets. While extrusion of the latter resulted in a degree of uniaxial orientation, this effect was virtually absent unless extrusion ratios in excess of approximately 10:1 were employed. But an extrusion ratio of 13:1 was sufficient to achieve alignment in a 15 to 20 vol % β-SiC/Al-2.5 Si alloy, liquid-phase hot-pressed without pre-alignment. The billet was 1-1/8 in. in diameter x 3 in. in length, canned in 1100 aluminum, and extruded at 530° to 560°C.[153]

Whisker Composite Properties

The achievable mechanical properties of whisker composites have become a subject of considerable disagreement[4] over the validity of presentation of laboratory data plotted together with properties calculated on the basis of the rule of mixtures. For discontinuous filaments the strength relationship has been severely questioned and experimental results do not support it. When interpreting laboratory data on whisker alloys it must be remembered, also, that nearly all data have been generated using small, substandard test coupons and test methods wholly, or in part, unacceptable to the engineering community.

The rule of mixtures as applied to composites is discussed in the subsequent section on mechanics. Equation 1 holds reasonably well for modulus (unidirectional composite) estimates, but the strength of discontinuous filament composites is not as easily estimated. Equations 48 through 53 give an approximation of this case.

The subsequent data are taken from the work on alumina[157,158] and silicon carbide[153,159] whiskers in metal matrices as representative of the most advanced mechanical property data yet obtained for this type of composite. The tensile strength of Al_2O_3/Al is given as a function of volume fraction alumina and temperature in Figure 93. Within the experimental limits of the data, composite strength increased with the amount of whisker reinforcement and was virtually insensitive to temperature to 500°C. The scatter in the strength values is typical of whisker-composite properties. The type of test performed was also not critical to the strength values obtained. The aspect ratio of the alumina whiskers (l/d) exceeded 1,000. The stress-strain curve for an instrumented specimen loaded to fracture with

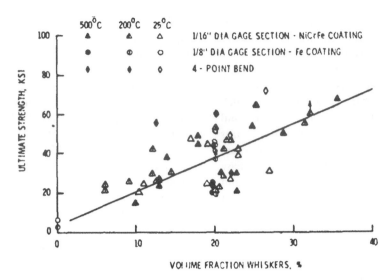

FIGURE 93.   Longitudinal tensile strength of Al_2O_3 whisker-Al composites at 25°, 200°, and 500°C.

two unloading cycles is given in Figure 94. Primary and secondary moduli were observed, the primary modulus being approximated by the rule of mixtures (or the Halpin-Tsai equation at l/d =∞). The secondary modulus was of the same order as the fiber contribution alone. The actual situation

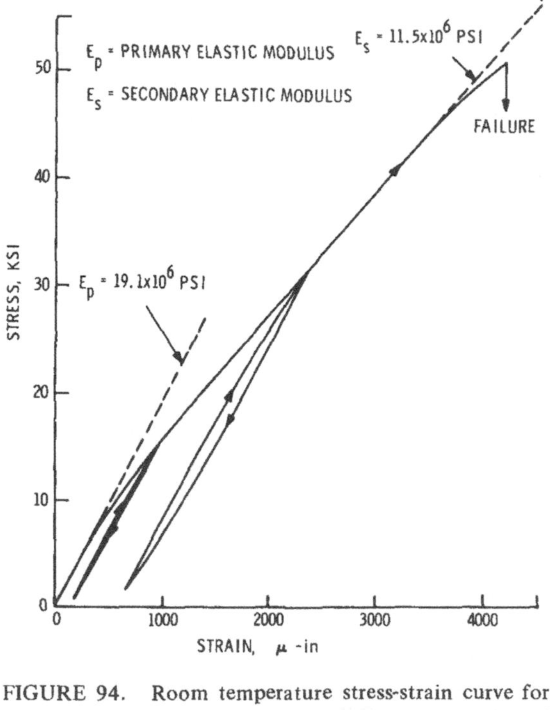

is probably much more complex than these results indicate. The hysteresis loops in composites have been previously reported.[131,166] Cyclic loading to a fixed tensile stress leads to a gradual increase in specimen length termed ratcheting.[157] This has been observed in the Si_3N_4/Ag system[167] but was earlier reported not to occur in Al_2O_3.[156] In that case, the cycling was at lower stresses (zero to 13,700 psi) and ratcheting was not detected. At zero to 25,500 psi, the specimen deformed about 0.1% before stabilizing after 1,500 cycles.

The longitudinal dynamic elastic modulus of Al_2O_3/Al as a function of whisker content is shown in Figure 95. Actually, the lower bound is better represented by a curved line going through the zero volume fraction at the matrix modulus of ten million psi. In the Halpin-Tsai representation, this would be the result for l/d = 1, and most values lie above this point and below the rule of mixtures value (l/d = ∞ for Halpin-Tsai equation). Most values are found to lie between the Halpin-Tsai prediction[168] of the moduli for l/d = 1 to l/d = 10. Since l/d approaches or exceeds 1,000, the results are somewhat lower than predicted. Factors such as matrix porosity and lack of complete fiber alignments are considered responsible[158] for the lower values. The elevated temperature moduli for three 20 vol % Al_2O_3/Al composites are given in Figure 96. The results have been compared to data obtained in the same manner for unreinforced 2024 aluminum and previous results on 2024 Al and Ti-6Al-4V.[169] The temperature dependence is

FIGURE 94.   Room temperature stress-strain curve for 20 vol % Al_2O_3 whisker-Al composite. Strain rate of 0.002 min^{-1}.

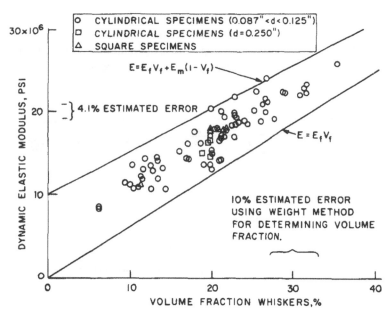

FIGURE 95. Longitudinal dynamic elastic modulus of Al_2O_3 whisker-Al composite as a function of whisker content.

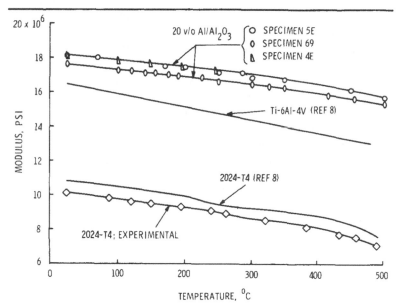

FIGURE 96. Elevated temperature dynamic longitudinal modulus of Al_2O_3 whisker-Al composite compared to several alloys.

seen to be quite similar in each case, with the Al_2O_3/Al composites holding up well to high temperatures. The creep-rupture properties as seen in Figure 97 are very good, giving an almost flat curve up to 500°C. A low creep rate of about 5 x 10^{-6} hr^{-1} was observed.[158] When failure occurred, it was abrupt with failure mainly by fiber

pull-out with matrix material adhering to the sides of the fibers. The results were similar to those reported by Kelly and Tyson[170] for W/Ag composites. No evidence of debonding at 500°C was found.

On the other hand, the fatigue behavior of these composites was very disappointing. The

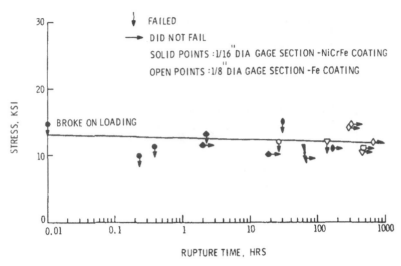

FIGURE 97.   Creep-rupture curve for 20 vol % Al$_2$O$_3$ whisker-Al composites at 500°C.

fatigue strength of a typical 20 vol % Al$_2$O$_3$/Al composite was substantially less than that of unreinforced 2024-T4 Al at all temperatures. This is shown in Figure 98. These results contrast with the higher fatigue strengths and lower high temperature strengths obtained with SiC reinforced aluminum.

The stress-strain behavior for a 25 vol % β-SiC/2024-T4 Al composite is shown in Figure 99.[159] The modulus is seen to increase from 10 x 10^{-6} to 21.6 x 10^{-6} for the composite, which is in reasonable agreement with the value predicted by Halpin[168] for l/d greater than 10. For β-SiC,

l/d was greater than 1000 before processing and ranged to as low as 70 after consolidation by extrusion. For α-SiC the as-furnished whiskers had an l/d of only about 100. The loading effect of repeated cycles resulted in a successive increase in yield strength with each cycle at least to the ten cycles tried.[159] The increase in tensile strength with increasing whisker content is shown in Figure 100 for a β-SiC/Al-2.5 Si composite, hot-pressed and extruded. Also, the lower values obtained using the lower l/d α-whiskers in the composites are shown. The values for hot-pressed billets were raised by more than one third through aging

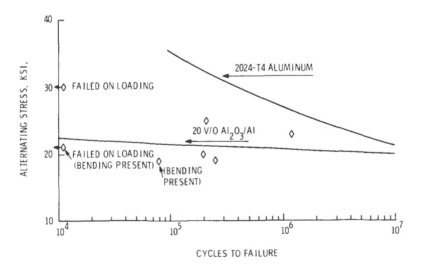

FIGURE 98.   Fatigue curves for 20 vol % Al$_2$O$_3$ whisker-Al compared to 2024 Al at room temperature under fully reversed loading.

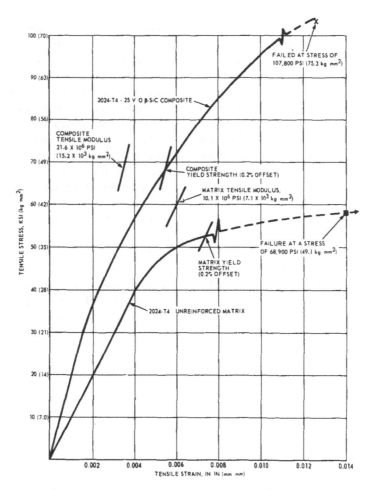

FIGURE 99. Stress-strain behavior of 25 vol % B-SiC whisker reinforced 2024-T4 Al composite.

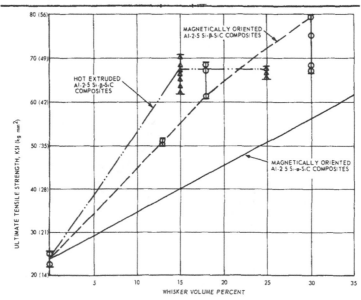

FIGURE 100. Ultimate tensile strength of Al-2.5 Si composites as a function of α- and β-SiC whisker concentration.

treatments. The tensile strengths at elevated temperatures were found to decline very rapidly as illustrated in Figure 101. The β-SiC whiskers were found to have fragmented considerably during extrusion. The strengths are still substantially higher than for the unreinforced Al-2.5 Si alloy. The rapid deterioration in high temperature strength of these SiC-reinforced alloys has been attributed[152] to the low aspect ratios of the SiC filaments compared to Al_2O_3 filaments where the reinforced aluminum alloys show no deterioration in strengths to 500°C. The off-axis tensile strengths of both Al_2O_3 and SiC-reinforced aluminum have been found to vary less than in the case of continuous fiber composites.[153] They exhibit no tendency to fail at low shear stress levels at orientations transverse to the whisker direction.

Very recent efforts[171,172] to achieve substantial reinforcement of aluminum alloys with β-SiC whiskers have been fairly encouraging in terms of compressive and tensile strengths obtained. High quality β-SiC whiskers fully separated from debris were consolidated with 2024 Al by hot pressing at 3,000 psi and 600°C. As-pressed densities were greater than 99% of theoretical. Both the composites and the base alloy specimens were heat treated and aged to achieve higher strengths. A solution treatment at 915°F (490°C) for 7 hr followed by artificial aging at 365°F (185°C) for 18 hr was found to be preferable to the commercial T-6 heat treatment for these hot-pressed specimens. The results for compressive properties are shown in Table 14. Considerable reinforcement was achieved in the direction of the reinforcement (0° to axis) and a reasonable improvement seen in random orientations. The transverse (90° to axis) results show that the matrix whisker bond is reasonably strong and at least the whiskers are not a source of weakness in

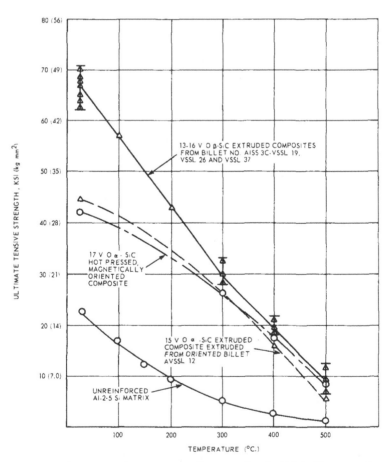

FIGURE 101. Degradation of tensile strength of Al-2.5 Si composites containing α- and β-SiC whiskers as a function of temperature.

TABLE 14
Compressive Properties of Unreinforced Matrix (2024 Al Alloy) and 15 and 25 v/o β-SiC Whisker Composites

| Test specimen no. | Whisker content | Whisker orientation to compressive axis | Yield strength at 0.2% offset | Compressive Properties | | Total strain to failure |
| | | | | Ultimate strength | Elastic modulus* | |
	v/o	degrees	kpsi	kpsi	Mpsi	in./in.
1L	–	–	47.8	87.4	11.3	0.1230
2L	–	–	49.2	82.4	11.0	0.1049
3L	–	–	48.2	87.4	10.8	0.1112
1T	–	–	50.6	88.0	11.6	0.1163
1L	15	0	155.2	270.5	21.7	0.0190
2L	15	0	144.6	249.4	19.1	0.0208
3L	15	0	170.6	264.4	22.8	0.0169
1T	15	90	46.9	79.3	14.5	0.0999
2T	15	90	50.0	81.6	13.9	0.1055
1L	25	0	182.3	193.9	24.8	0.0105
2L	25	0	¡00	196.5	28.3	0.0084
3L	25	0	192.4	214.9	26.7	0.0100
1T	25	90	52.0	91.8	16.6	0.0691
2T	25	90	50.6	88.9	15.1	0.0939
1	15	Random	81.8	143.4	13.8	0.0253
2	15	Random	83.3	140.3	16.2	0.0223
1	25	Random	84.8	156.8	17.4	0.0233
2	25	Random	84.0	162.3	16.5	0.0271

*From final loading curve

the composite since the results are similar to those in an unreinforced matrix. The tensile properties are given in Table 15. Here again the reinforcement achieved at 0° to the axis of the whiskers is substantial. Random orientation also achieves some strengthening while the transverse tensile strengths (90° to axis) approximate the matrix strength. The compressive strengths are significantly greater than the tensile strengths in each case, while the moduli are the same. The tensile value of 171 ksi for a 25 vol % B-SiC/2024 Al composite is quite promising. Unfortunately, attempts at extrusion of rods of 20 vol % β-SiC62024 Al yielded poor strengths, apparently due to restricted matrix flow by the longer, stronger β-SiC whisker compared to α-SiC whiskers.[171] Difficulties in hot pressing 40 vol % β-SiC/2024 Al composites at 8,000 psi and 640°C produced specimens of less than 90% of theoretical density. For the moment the whisker-

reinforcement in this system seems to be limited to lower volume loadings.

Flexural fatigue testing was conducted on a series of β-SiC reinforced composites with 7075-T6 Al and Al-2.5 Si alloys at 1,380 cycles per minute. The results are given in Figures 102 and 103.[159] The comparison values for unreinforced 7075-T6 Al and Al-2.5 Si show a substantial increase in fatigue strength in the whisker composite structures. A comparison of these data with those for Al_2O_3/Al composites appears to indicate that high aspect ratios do not necessarily improve fatigue properties. The fatigue fractures occurred by brittle failure. For all alloys tested, 2024-T4, 6061-T4, 7075-T6, and Al-2.5 SiC, the fatigue strength of the composite was always much better than for the matrix. This is one of the more attractive findings of current investigations which should encourage further research on discontinuous filament composites.

TABLE 15

Tensile Properties of 2024 Al Matrix and 25 v/o β-SiC Whisker Composites

Test sample no.	Whisker content	Whisker orientation to tensile axis	Yield strength at 0.2% offset	Ultimate strength	Elastic modulus	Total strain to failure
	v/o	degrees	kpsi	kpsi	Mpsi	in./in.
1	25	0	—	>152.3[a]	24.9	0.0081
2	25	0	—	>164.1[a]	23.0	0.0084
3	25	0	—	170.8	25.2	0.0093
1	25	90	48.8	49.2	16.7	0.0050
2	25	90	—	50.0	17.4	0.0035
3	25	90	52.2	56.3	17.2	0.0077
1	25	Random	67.2	92.4	17.6	0.0126
2	25	Random	62.6	88.9	17.3	0.0131
3	25	Random	68.4	93.4	17.4	0.0124
1	0	—	43.2	58.3	11.0	0.0624
2	0	—	44.8	61.6	12.0	0.1167
3	0	—	43.1	57.4	9.9	0.0701

(a) Failure in grips

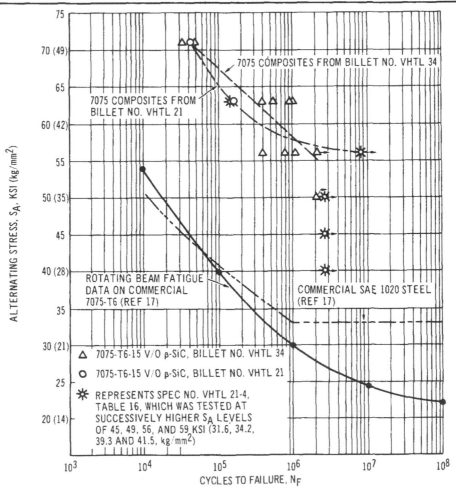

FIGURE 102.   S-N curves for β-SiC whisker-7075 T6 Al composites.

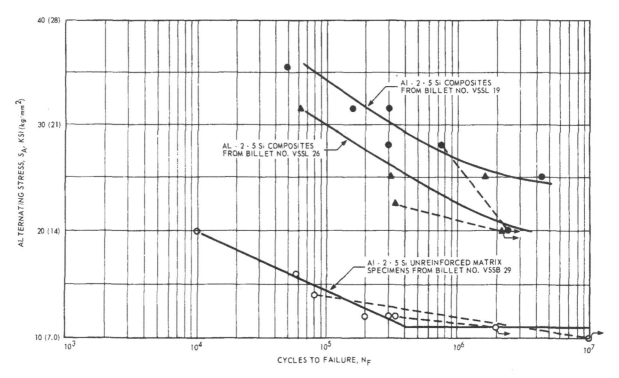

FIGURE 103.   S-N curves for β-SiC whisker-Al-2.5 Si composites.

Chapter 7

TENSILE PROPERTIES

Stress-Strain Behavior

The tensile properties of state-of-the-art boron reinforced aluminum and titanium are summarized in Table 16. Typical stress-strain diagrams for unidirectionally reinforced B/Al composites tested parallel (0°) and perpendicular (90°) to the direction of reinforcement are shown in Figure 104. There are several interesting features of these stress-strain curves which deserve special mention.

If the 0° curve is critically examined, three distinct regions are noted. With the initial application of load, both phases deform elastically with the rule-of-mixture relation giving a fairly accurate prediction of modulus. Eventually, the yield strength of the matrix is exceeded and the matrix begins to flow plastically. As a result, the matrix contribution to composite stiffness is substantially reduced. Composite stiffness at a given strain is then determined by the weighted average of the modulus of the reinforcement and the instantaneous strain hardening rate of the matrix, $\frac{d\sigma}{d\epsilon}$. In the case of B/Al, $\frac{d\sigma}{d\epsilon}$ for aluminum is negligibly small compared with the 59×10^6 psi modulus of boron filament. The slope of the stress-strain curve in this region is referred to as the system's secondary modulus and is generally 70 to 90% of the initial modulus.[38,173] The stress-strain behavior in this second region is, of course, neither elastic nor linear. The system experiences permanent deformation by reason of matrix flow and breakage of severely weakened filaments, and the strain hard-

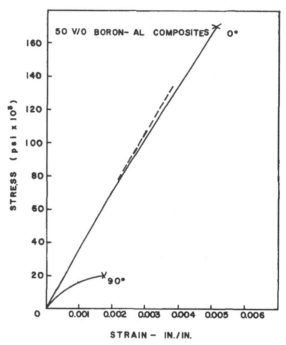

FIGURE 104. Typical stress-strain curves for 50 vol % unidirectional boron/aluminum composites tested parallel (0°) and perpendicular (90°) to the filament.

ening rate of the matrix is not necessarily constant, although changes in matrix hardening rate have little influence on the composite stress-strain behavior in this region. This second stage continues until filament breakage is encountered, whereupon the slope of the stress-strain diagram is

TABLE 16
Tensile Properties of Boron/Aluminum and Boron/Titanium Composites

		50 v/o Boron-Aluminum		50 v/o Boron-Titanium	
		Base(psi)	Weight Normalized(in.)	Base(psi)	Weight Normalized(in.)
Ultimate Tensile Strength	0°	165,000	1.78×10^6	155,000	1.22×10^6
	90°	15,000	0.16×10^6	30,000	0.24×10^6
Tensile Modulus	0°	33.4×10^6	356×10^6	38.0×10^6	300×10^6
	90°	21.0×10^6	224×10^6	28.0×10^6	220×10^6

again observed to decrease (stage III), eventually resulting in composite fracture. If the yield strain of the matrix exceeds the fracture strain of the brittle filament, stage II-type behavior will not be observed. The transition from stage I to stage II depends on the yield strength (or strain) of the matrix and the magnitude of residual consolidation stresses.[38] In the case of "ductile" metal wire reinforcement, stage III is extended because both phases are capable of deforming plastically, causing failure to be postponed. These various stages are schematically shown in Figure 105.[174]

A rule-of-mixture type prediction of composite tensile strength will require knowledge of the stress-strain behavior of each component under the conditions it would experience in the composite.[174] This includes such subtle points as matrix grain size, impurity distribution, consolidation induced defects, the interdiffusion of components, and the presence of reaction products. Filament strength and strain-to-failure must reflect the chemical and/or mechanical degradation resulting from consolidation and forming operations. Since filament tensile strength is sensitive to gage length, it is very conservative to use filament strengths measured on 1-inch gage lengths in composites where the critical load transfer

length is generally less than 0.1 inch.[2,175] The influence of closely spaced stiff and strong filaments on the flow stress of the matrix and the possibility of chemical change as a result of reaction with the filaments must be considered.[38,174] Residual consolidation stresses will influence the response of both phases to applied loads.[176] Using the isostrain criterion, the contribution of each phase to the composite 0° tensile strength will depend on the strength developed in the two phases at the low fracture strain of the brittle filaments. Composite failure strains are generally between 0.3 and 0.6%. Anything which reduces the strain-to-failure of the filaments will be directly evidenced in reduced composite 0° tensile strength.

Residual Stresses

The aforementioned residual consolidation stresses arise because of the thermal expansion mismatch of the two components. For instance, the thermal expansion coefficient of boron is 2.8 micro in./in./°F while 6061 aluminum is 13.1 micro in./in./°F and titanium 6Al-4V is 4.7 micro in./in./°F.[143] Since consolidation of filament and metal takes place at relatively high temperatures (1000°F (538°C) for B/Al and 1500°F (816°C) for B/Ti), these differences in expansion coefficient will result in the formation of longitudinal and radial residual compressive stresses on the filaments and corresponding tensile stresses in the matrix. Quantitative estimates of the magnitude of these residual stresses have been made, but these estimates are complicated by the relaxation of the matrix both during cooling from the consolidation temperature and at room temperature after consolidation.[176]

Calculation of the matrix residual stress, using the equation[177]

$$\sigma_m^r = \frac{E_m \Delta\alpha\Delta T}{\frac{E_m V_m}{E_f V_f} + 1} \qquad (1)$$

for a 50 vol % B/Al composite, assuming, as an example, an effective ΔT of 300°F (149°C), results in a $\tau_m^r = 26,000$ psi. This simple calculation has assumed a purely elastic response from the matrix, which is not to be expected in this case, but it does give an indication of the magnitude of the residual stresses which are generated.

In most cases one finds that matrix yielding

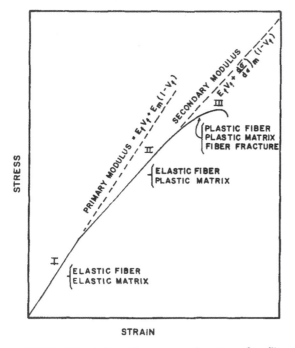

FIGURE 105. Schematic stress-strain curve for filamentary reinforced metals showing three regions.

takes place during cooling so that in the absence of relaxation the matrix can exhibit no elastic deformation when the composite is loaded in tension. Since elastic behavior is observed one must conclude that matrix relaxation has taken place. If the filament-matrix bond strength is inadequate, cooling will result in slippage along this boundary instead of causing matrix strain hardening, and the magnitude of the resultant residual stresses will be reduced. This condition, however, has never been observed in systems of practical interest.

The influence of these residual stresses on composite behavior can take two forms. First, since the filaments are initially loaded compressively, the composite failure strain in tension is higher than would be predicted if residual stresses were ignored. As the composite is loaded, the filaments are initially unloaded compressively and subsequently loaded in tension. Effectively, the filament strain-to-failure is increased. Since, in general, the filament tensile failure strain determines composite failure strain and, thereby, the tensile strength of the composites, this increased strain-to-failure will result in higher composite tensile strengths. The strain hardening matrix can make a larger contribution to composite strength. In addition, the initial load carrying capacity of the matrix is considerably higher than expected as a result of prior strain hardening. These effects are shown schematically in Figure 106.[177] Unfortunately, these triaxial matrix tensile stresses will add to the problem of matrix embrittlement, particularly for cross- or angle-plied composites where extremely high tensile stresses can be generated and where lateral accommodation of the stresses is impossible except in the short transverse direction. Low composite failure strains have been observed for these cross- and angle-plied materials using normal consolidation processes. A technique which has helped alleviate this matrix triaxial tensile stress state is a low temperature quench. Cooling the composites below room temperature causes further matrix flow to help accommodate the thermal expansion mismatch. Subsequent heating to room temperature causes a relaxation of the residual stresses and an increase in composite failure strain.[178]

Transverse Tensile Properties

There is a striking difference between the 0° and 90° tensile properties of these unidirectionally reinforced materials (Table 16 and Figure 104).

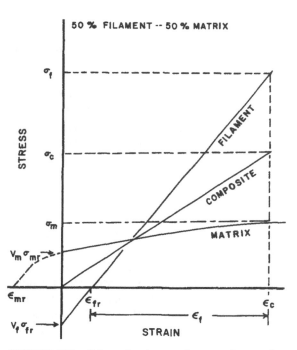

FIGURE 106. Schematic stress-strain curves for matrix, filament and composite demonstrating the influence of residual stresses.

The tensile modulus, strength, and ductility are all lower for transverse tensile tests than for longitudinal tensile tests. In fact, there is a 10:1 difference in strength and a 1.5:1 difference in modulus.[123,179] The lower moduli and strength are, in part, due to the fact that the isostrain criterion no longer applies. That is, the matrix is free to flow nearly independently of the filaments. Under these conditions it becomes more difficult to predict composite stiffness and strength. If, however, the filaments are well bonded to the matrix and free of defects, the transverse strength should approach or exceed the strength of the bulk matrix alloy.[102] Unfortunately, these strength levels are not usually attained in either boron reinforced aluminum or titanium alloys (Tables 16 and 17). The explanation for low transverse strength and ductility seems related to the transverse strength of boron filament. Diametral compression tests on 0.5 inch long 4-mil boron filaments have shown that there is a large distribution of filament strengths and that this distribution is skewed sharply to the lower strength levels with a mode of approximately 30,000 psi.[123] If filaments split in transverse tensile tests of composites at these same low stress levels, the load carrying cross section would be

TABLE 17

Mechanical Properties of 6061 Aluminum and Titanium 6Al-4V

		σ_y(psi)	σUTS(psi)	ϵ(%)	τ(psi)	E(10^6 psi)	G(10^6 psi)
Aluminum-6061	T-0	8,000	18,000	25	12,000	10.0	3.75
	T-6	40,000	45,000	12	30,000	10.0	3.75
Titanium-6Al-4V	A*	120,000	135,000	11	78,000	16.5	6.1
	ST&A**	150,000	170,000	7	95,000	16.5	6.1

* Annealed @1325° F for one hour and furnace cooled at less than 50° F/hr to 800° F.

** Solution treated @1750° F for 30 minutes, air cooled, aged @1000° F for six hours.

sharply reduced and severe stress concentrations would be introduced which would account for the observed low composite strength and ductility.

Examination of transverse tensile fracture surfaces clearly establishes the degrading influence of filament splitting. For well bonded composites with moderately strong matrices, fracture surface inspection indicates that nearly all of the filaments have been longitudinally split (Figure 107).[123,178] Figure 108 shows the influence of

matrix strength on filament splitting.[178] For a weak matrix (Figure 108A) transverse failure is controlled by matrix strength and failure takes place predominantly through the matrix. If the matrix in these same composites is solution treated and aged so that it can introduce higher loads into

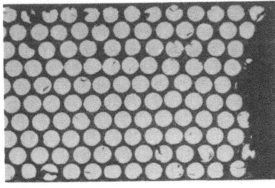

FIGURE 108A.

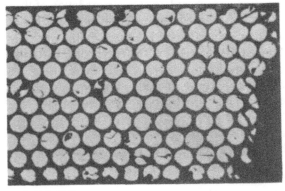

FIGURE 108B.

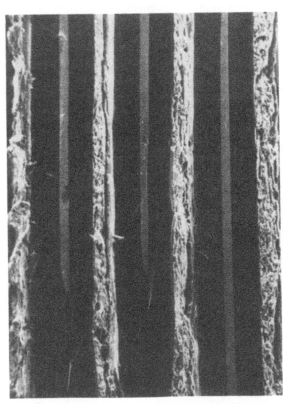

FIGURE 107. Transverse tensile fracture surface of a boron/aluminum composite showing split boron filaments.

FIGURE 108. Transverse tensile fracture surface of boron/aluminum composites showing the influence of matrix strength on filament splitting and fracture. A. Annealed condition. Matrix stength controlled failure. B. Solution treated and aged condition. Filament transverse strength controlled.

the filaments, transverse failure is controlled by filament splitting and failure occurs predominantly through the filaments (Figure 108B). Additional evidence to support these conclusions is presented in Figure 109, where the transverse strength for two aluminum alloy heat treatment conditions is plotted as a function of vol % boron.[178] Since the transverse strength of annealed specimens is found to be independent of boron content and to have strengths of approximately 80% of that of annealed 6061 aluminum, one can conclude that the boron filaments are contributing a transverse load carrying capacity nearly equivalent to the load carrying capacity of an equivalent volume of aluminum. The slight strength reduction can be related to structural imperfections in composites and the splitting of very weak boron filaments. Quantitative examination of the fracture surfaces indicates that in all cases the relative area of split filaments on the fracture surface is less than the vol % boron in the composites. Failure is controlled by matrix properties. This contrasts with composites where the matrix has been heat treated and aged to a T-6 condition. The matrix is now substantially stronger than the majority of the boron filaments (i.e., 45,000 psi for T-6 6061 Al compared with a 30,000 psi mode for boron filament splitting). Here the transverse strength is found to be a fairly sensitive function of boron content as would be expected if filament splitting controls the composite failure. Generally, the

relative area of split filaments on the fracture surface is found to be larger than the vol % boron, which clearly indicates that the filaments offered the least resistance to crack propagation and probably control composite failure. The higher transverse strengths of the heat treated system at all reinforcement levels reflect the higher strength and stress concentration accommodation ability of the matrix.

Composites have been fabricated with aluminum alloys with strengths ranging from 17,000 psi to 39,000 psi.[123] The transverse strengths of these composites were found to be independent of matrix strength (Figure 110), although the magnitude of the transverse strength (13,000 psi) is considerably lower than the equivalent strength reported in Figure 109 (22,000 psi). This difference in transverse strength may have resulted from a difference in initial filament quality or consolidation technique. Examination of the fracture surfaces of the test specimens reported in Figure 110 all indicated extensive filament splitting. In the case of aluminum matrix composites, filament splitting has only been observed at, or in the immediate vicinity of, the composite fracture surface. These observations reflect the susceptibility of aluminum alloys to the severe stress concentrations associated with split boron filaments and their low strength. In contrast, extensive filament splitting throughout the entire gage length has been observed for B/Ti composites

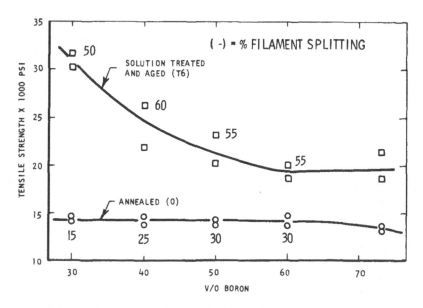

FIGURE 109.   Transverse tensile strength of boron/6061 aluminum composites as a function of filament content for the matrix in the T-O and T-6 conditions.

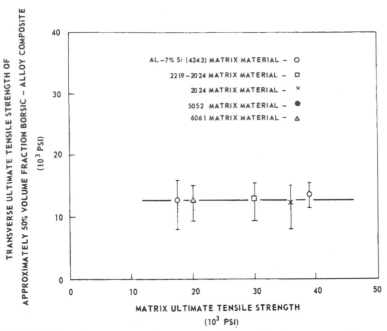

FIGURE 110. Transverse tensile strength of boron/aluminum composites as a function of matrix strength.

(Figure 111).[180] This figure shows that every filament in the entire gage section has split normal to the applied load. Apparently, the titanium matrix has been able to strain harden and redistribute the load and appears to be less susceptible to the stress concentrations associated with split filaments. A direct comparison of Figures 108 and 111 is not valid, however, because of the large difference in boron filament content. The extensive filament splitting shown in Figure 111 may have been facilitated by the consolidation process (continuous roll bonding), but no split filaments were observed in the as-fabricated specimen.

The aluminum-titanium comparison which has been developed above is not consistent with the observed transverse tensile strengths in these composite systems. If the filaments were assumed to make no contribution to transverse strength (i.e., filaments replaced by cylindrical voids), a 50 vol % composite ought to exhibit a transverse tensile strength of approximately one half the tensile strength of the bulk matrix. While this is a simplified view, it does provide a basis of comparison. For annealed titanium 6Al-4V and T-6 6061 aluminum these predicted strengths would be 70,000 psi and 22,000 psi, respectively. For aluminum composites, transverse tensile strengths in excess of 30,000 psi have been achieved while

for titanium composites the highest reported transverse strengths reported are around 60,000 psi.[180] This would imply that there has been more success in fabricating aluminum matrix composites than titanium matrix composites. This conclusion is supported by the low 0° tensile strengths and the larger spread associated with a given property measurement in titanium composites.

Improvement of the transverse properties of unidirectionally reinforced B/Al can be accomplished by development of improved filaments, by third phase additions, or by heat treatment of the matrix. An alternate solution is to use cross- or angle-plied B/Al where substantial off-axis loads are anticipated. The use of cross- or angle-plies (1) imposes a reduced upper limit on the volume fraction of boron filament since interpenetration of adjacent boron layers becomes impossible, (2) increases the probability of introducing filament defects during consolidation, and (3) introduces complex shear coupling forces between adjacent boron layers. None of these features in themselves would eliminate cross- or angle-plied composites from active consideration but they do introduce complexities which, in most cases, could be avoided if the transverse strength of unidirectional composites were increased to 40-50,000 psi. The first of the solutions, improved filaments, is being

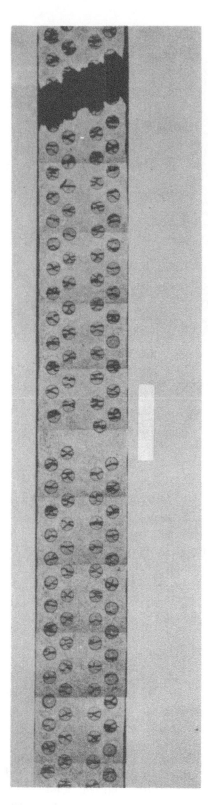

FIGURE 111. Transverse tensile failure of boron/titanium composite showing extensive filament splitting normal to the applied load.

pursued by the development of larger diameter (8 mil) boron filament, by the use of different boron filament substrate (carbon), and by the development of additional varieties of filament (SiC, Al_2O_3, and graphite). Filament development is usually, however, a long and costly process. The second solution, third phase additions, shows great promise. Composites have been fabricated and tested with minor (5-15 v/o) 90° stainless steel wire additions and with layers of titanium foil which demonstrate substantial improvements in transverse properties without materially affecting longitudinal properties.[181] Matrix heat treatment, in addition to modifying the matrix strength and its susceptibility to stress concentrations, can also be used to change the residual stress state in the composites. Because of the higher thermal expansion coefficient of aluminum and titanium relative to boron, rapid cooling will result in residual radial compressive stresses on the boron filament which may increase composite transverse strength and ductility in those systems currently limited by filament splitting.[123]

Filament-Load Angle

Typical variations of tensile strength and modulus with filament-load angle for unidirectional 50 vol % boron rereinforced aluminum and titanium composites are shown in Figures 112 and 113.[36,38,123,179] The variation of the fracture surface with filament-load angle is shown in Figure 114.[36] The sensitivity of composite tensile strength to filament-load angle highlights the requirements for accurate test methods. At small angles (1° to 2°) composite failure is primarily determined by filament strength, but this effect is lost almost immediately and the shear strength of the matrix soon determines composites failure.[174,182-184]

$$\sigma_c = \tau_m/\sin\theta \cos\theta \qquad (2)$$

The transition from filamentary to matrix shear controlled failure has been predicted to take place at an angle determined by the ratio of the matrix shear strength to the 0° composite tensile strength.[151]

$$\tan\theta_{crit} = \tau_m/\sigma_c \qquad (3)$$

For the B/Al composite data shown in Figure 112, using the bulk shear strength of annealed 6061 aluminum, this critical angle is calculated to be 4°. Since almost a third of the B/Al composites 0°

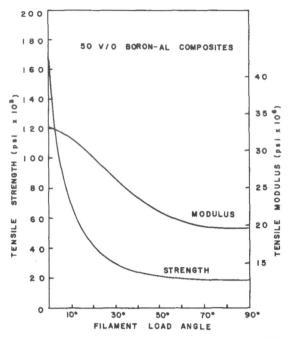

FIGURE 112. Tensile strength and modulus of 50 vol% unidirectional boron/6061 aluminum composites as a function of the filament-load angle.

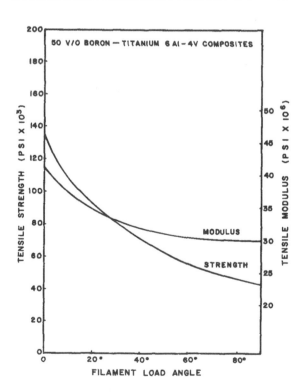

FIGURE 113. Tensile strength and modulus of 50 vol% unidirectional boron/titanium 6 Al-4V composites as a function of the filament-load angle.

FIGURE 114. Fracture surfaces of unidirectional Borsic/6061 aluminum composites tested at various filament-load angles.

tensile strength is lost within the first 5°,[185] the extent of filamentary controlled failure is considerably less than that predicted. This same critical angle for B/Ti composites is calculated to be nearly 30° which is evidently far from reality. Shearing of the matrix parallel to the filaments controls composite failure up to an angle of 50° to 60° where matrix plane strain failure becomes dominant.[184]

$$\sigma_c = (\sigma_{uts})_m / \sin^2\theta \qquad (4)$$

Although composite tensile failure is much more complex than has been implied, the equations expressed above yield surprisingly accurate predictions of composite strength as is shown in Figure 115. The matrix shear and tensile strengths used in the comparisons in Figure 115 are handbook values. Measurements of shear and tensile strengths in thin brazed joints have indicated substantial increases over the corresponding strengths of the bulk metals due to constraints of the more rigid joint shoulders.[118] Since similar boundary conditions exist in fila-

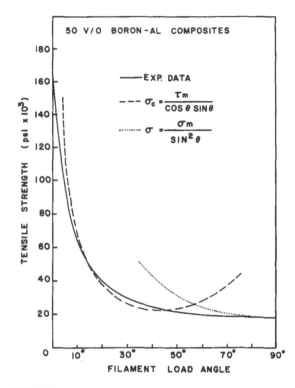

FIGURE 115. A comparison of experimental data of the variation of composite strength with filament-load angle with theory.

mentary reinforced metals, the use of handbook values of shear and tensile strengths for the comparisons cited above is not realistic. Taking this constraint into consideration, the matrix shear strength might be increased by as much as 50% and the matrix tensile strength increased by 10 to 20%.[174] If these adjusted strengths had been used in Figure 115, the agreement with composite tensile data would not have been nearly as good. The additional refinement of determining matrix properties in composites would be important but would not materially add to the accuracy of these predictions.

A minimum is sometimes observed in the ultimate tensile strength at 45° because slip can take place on planes of maximum resolved shear stress parallel to the reinforcement.[184] From 45° to 90° there may be a slight increase in strength as the matrix is again required to flow around filaments. This 45° minimum has been observed in several systems (Figure 116),[179] particularly for thin gage sections, but is not usually very pronounced when it is observed. Instead of a slight increase, a slight decrease is commonly observed as

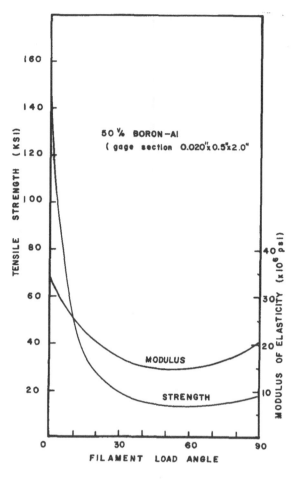

FIGURE 116. Tensile strength and modulus of 50 vol % unidirectional boron/6061 aluminum composites as a function of filament-load angle showing a minimum at 45°

the filament-load angle is increased from 45° to 90°. In this case it is always possible to find a surface where shear can take place at 45° to the tensile axis (maximum resolved shear stress) but which is not normal to the major face of the gage section. The physical extent of this surface decreases continuously as the filament-load angle is increased and increases with increased filament content.[184] Figure 117 shows the variation of tensile strength with load angle for various filament configurations for 50 vol % boron-aluminum composites.[2,179] The ability to tailor-make the biaxial properties of these composites is demonstrated. The corresponding variation of modulus with load angle is shown in Figure 118.[2,179] Notice that the modulus of the cross- and angle-ply composites is shown to be insensitive to load angle.

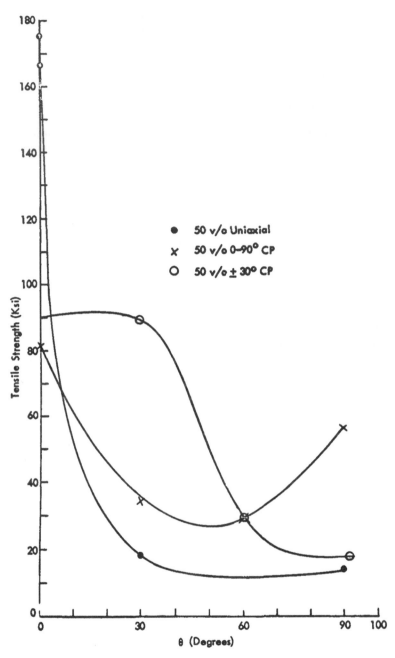

FIGURE 117.   Tensile strength of unidirectional, cross plied and angle plied 50 vol % boron/6061 aluminum composites as a function of filament-load angle.

Reinforcement Content

The variation of 0° tensile strength and modulus with boron content for unidirectionally reinforced B/Al composites is shown in Figure 119.[153,178,179] These values are in good agreement with rule of mixture type predictions. The decreased strengthening rate at high filament loadings (>60 v/o) is related to consolidation difficulties, severe matrix constraint, and to the higher frequency of filament-filament contact.[186] The maximum volume fraction obtainable with equal size cylindrical filaments can be shown to be 0.906 but, practically, filament content is limited to something less than 0.80. If filamentary "reinforcement" were going to result in higher composite strength than the bulk strain hardened matrix alone, the following inequality would apply:

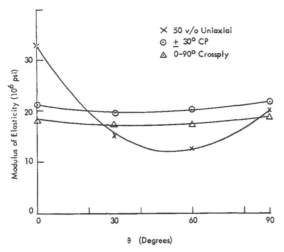

FIGURE 118. Tensile modulus of unidirectional, cross plied and angle plied 50 vol % boron/6061 aluminum composites as a function of filament-load angle.

$$(\sigma_f v_f + \sigma_m'(1 - v_f) \geq \sigma_{uts})_m \qquad (5)$$

where σ_m' is the flow stress of the bulk matrix at the fracture strain of the filaments.[174] A critical filament content (V_{crit}) therefore is defined for reinforcement as:[174]

$$V_{crit} = \frac{(\sigma_{uts})_m - \sigma_m'}{\sigma_f - \sigma_m'} \qquad (6)$$

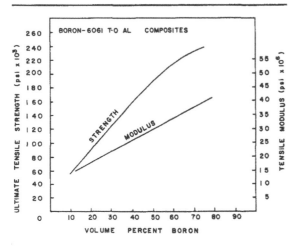

FIGURE 119. Tensile strength and modulus of unidirectional boron/6061 aluminum composites as a function of boron content.

Note the sensitivity of this expression to matrix strain hardening. Using the data in Tables 16 and 17, the critical boron filament content in 6061 aluminum alloys is 1 to 2 vol % and in titanium 6Al-4V alloys 3 to 5 vol %. In comparison, this critical value for tungsten wire reinforced nickel has been reported to be approximately 20 vol %.[174] These relationships are graphically displayed in Figure 120.[174] For extremely low filament volume fractions the fully strain hardened matrix containing cracked non-load carrying filaments would be stronger than reinforced systems assumed to fail at the low fracture strain of the filaments. This prediction $[\sigma_c = (\sigma_{uts})_m (1 - V_f)]$ does not consider the influence of stress concentrations associated with cracked filaments, a reasonable approximation. As discussed previously, these predictions would be altered if the flow stress of the metal in the composite were used (i.e., σ_m' would be increased). Furthermore, if ductile wires were used as reinforcements, there would be reinforcement at all volume fractions.[174]

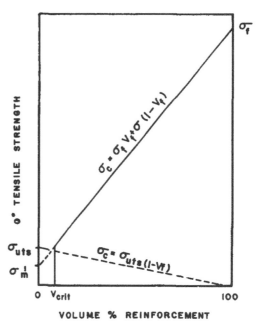

FIGURE 120. Predicted variation of 0° tensile strength with reinforcement content for filamentary reinforced systems.

Chapter 8

COMPRESSION

The ultimate strength of composites tested in compression has been found to equal or exceed their ultimate tensile strengths. The modulus of elasticity is nearly identical in tension and compression. Typical stress-strain diagrams for 0° and 90° specimens tested in compression are shown in Figure 121.[187]

The compression of 0.20 in. x 0.25 in. x 0.75 in. specimens of unidirectionally reinforced 50 vol % Borsic/Al parallel to the filaments (Figure 121) resulted in a modulus of 34 x 10⁶ psi and a compressive strength of 297,000 psi with failure

occurring as a result of the "brooming" of one end of the specimen.[36] In comparison, a similar sized specimen tested at 90° to the filaments resulted in a modulus of 20 x 10⁶ psi and an ultimate compressive strength of 37,000 psi. Failure of this specimen set in at approximately 6% strain and was caused by a matrix shear failure at roughly 45° to the load axis (Figure 122).[187] These transverse compressive strengths and ductilities are a dramatic, but perhaps expected, improvement over the equivalent tensile properties previously discussed. A similar specimen tested at 600°F

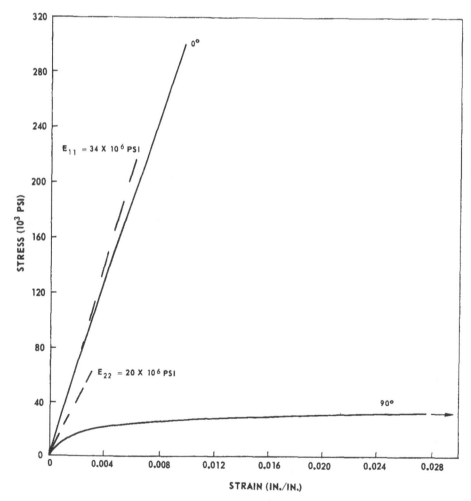

FIGURE 121. Typical compressive stress-strain curves for 50 vol % unidirectional boron/ aluminum composites tested parallel (0°) and perpendicular (90°) to the filament.

FIGURE 122. Matrix shear failure in a 50 vol % boron/6061 aluminum unidirectional composite tested perpendicular to the filament.

The variation of compressive strength of boron-aluminum composites with boron content is shown in Figure 123.[179] Note that the compressive strengths shown in Figure 123 are considerably less than those discussed previously. The data in Figure 123 were acquired on 0.020-in. and 0.080-in. sheet specimens while the higher strengths were acquired from the aforementioned columns. These different test procedures partially account for the differences in test results. Figure 124 shows the variation of compressive strength with filament-load angle for 25 and 50 vol % boron-aluminum composites.[179,188] Note here again that the 0° compressive strength is very much more sensitive to boron content than the 90° strength, the transverse compressive strength being determined primarily by the shear strength of the matrix. In compression the matrix not only transfers the loads into the filaments and around broken filaments but also must stabilize the filaments to prevent their buckling.[185,188] The failure mode for 0° specimens is usually "brooming" of the ends rather than failure in the gage section, which leads one to believe that an improvement in test technique should lead to even higher measured compressive strengths.

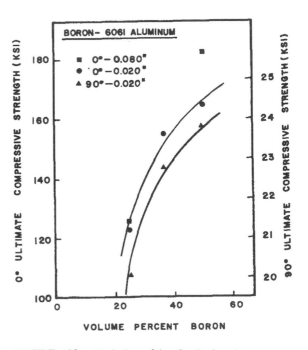

FIGURE 123. Variation of longitudinal and transverse compressive strength with boron content for unidirectional boron/aluminum composites.

resulted in a 9,500 psi ultimate compressive strength with the same type of matrix shear failure.[187] Since transverse failure in compression is linked to matrix shear, metallurgical treatments to increase matrix shear strength will be reflected in higher transverse compressive strengths.

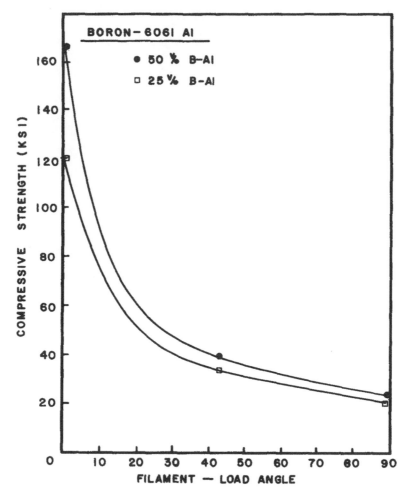

FIGURE 124. Variation of compressive strength of 25 and 50 vol % boron/6061 aluminum composites with filament-load angle.

Chapter 9

FATIGUE

The fatigue properties of filamentary reinforced metals are generally found to be superior to their monolithic counterparts. Because of the additional variables introduced by reason of the two phase anisotropic character of these materials, the prediction of fatigue properties and the analysis of fatigue test data become a very difficult task. This process is further complicated by the variety of loading schemes (flexural, axial, filament-load angle, and various combinations thereof) and the different definitions of fatigue failure. Metal-lurgical variables such as the strength, modulus, work hardening characteristics and notch sensitivity of the matrix, properties and physical form of chemical reaction products, effective strength and modulus of the filament-matrix bond, the strength, modulus, and stress concentration sensitivity of the filament, and the presence of preexisting or consolidation-induced defects all influence the response of these materials to cyclic loads. Some of the more significant features which have been identified will be discussed in the following paragraphs.

Fatigue Curves

Typical tension-tension fatigue curves for 25 vol % B/6061 Al unidirectional and angle-plied composites are shown in Figure 125.[186,189] The ratio of minimum to maximum stress ("R" ratio) in these tests was 0.2. Note the flatness of these curves. The ratio of the fatigue strength at 10^7 cycles to the ultimate tensile strength is given to the right of each curve for reference. Although there is no systematic variation of this ratio, the

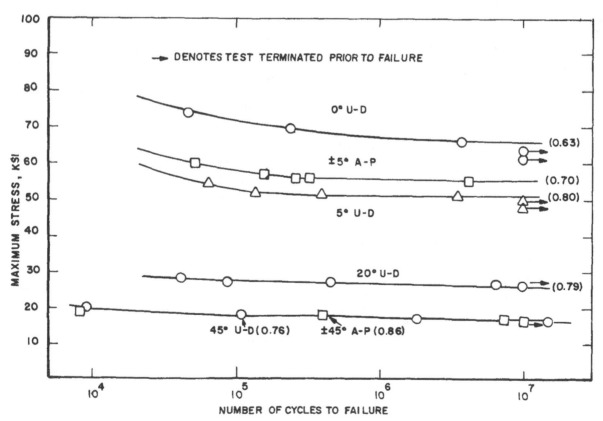

FIGURE 125. Fatigue curves for 25 vol % boron/6061 aluminum composites with various filament and filament-load configurations.

fatigue strength at 10^7 cycles is approximately 80% of the tensile strength in all cases. The fatigue strength of T-O 6061 aluminum tested under similar conditions is about 16,000 psi at 10^7 cycles while the ultimate tensile strength of similarly diffusion-bonded 6061 aluminum foil was found to be 22,000 psi. Using these figures, the ratio of fatigue strength to tensile strength for the 6061 aluminum matrix is about 0.73.[183,189] The similarity between the composite and matrix ratios leads one to suggest that composite fatigue failure may be controlled by the matrix. Comparing the unidirectional and ±5°, ±45° angle-plied composite curves, the stress concentrations and shear coupling forces introduced at filament crossover points are not reflected in reduced fatigue strengths, at least at this low boron volume fraction. The variation of fatigue strength with boron content for B/6061 Al composites is shown in Figure 126.[179,183,189] The "R" ratio for these tests was maintained at 0.2.

Combined stress (tension-bending) fatigue tests have been performed on 22 vol % boron composites contained in a matrix which was 55% 1100 and 45% 2024 aluminum.[24] The ratio of alternating stress to mean axial tensile stress was maintained at 0.95. The fatigue strength at 10^7 cycles both at room temperatures and at 500°F (260°C) was found to be approximately 46,000 psi. The ratio of fatigue strength at 10^7 cycles to ultimate tensile strength was found to be about 0.50 at both temperatures. In comparison, for 0° reverse bending flexural fatigue of 42 vol % B/2024 Al composites the fatigue strength has been found to be temperature insensitive up to 250°F (121°C) (107,000 psi) while at 500°F (260°C) the fatigue strength has been found to be reduced 42% (62,000 psi).[173] Reverse bending fatigue failure, at least at 500°F (260°C), would seem to be matrix-controlled.

Axial tension-tension fatigue tests have been performed on B/6061 Al composites with similar boron contents but different 0° ultimate tensile strengths (185,000 psi vs. 136,000 psi) using a ratio of alternating stress to mean stress ("A" ratio)* of 0.8.[2,36,173] The fatigue strengths at 10^7 cycles for these composites were found to be 90,000 psi and 85,000 psi, respectively. Despite a substantial difference in tensile strength the

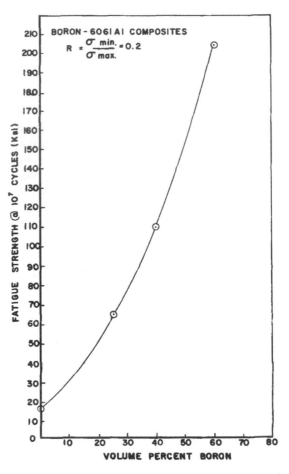

FIGURE 126. Variation of fatigue strength at 10^7 cycles with boron content for boron/6061 aluminum unidirectional composites tested in tension-tension fatigue.

fatigue strengths were very similar. This would imply that the current practice of optimizing tensile strength will not necessarily result in optimum fatigue strength. These differences most likely reflect the different failure mechanisms in tension and fatigue. The influence of maximum stress and "A" ratio on fatigue life is shown in Figure 127 for 46 vol % boron-6061 aluminum composites.[36] The average tensile strength for these specimens was 136,000 psi.

Filament-Load Angle

Figure 128 shows the variation of the fatigue strength at 10^7 cycles with filament-load angle for unidirectional, ±5° and ±45° angle-plied 25 vol % B/6061 Al composites.[186,189] As the filament-

*A = 1 – R/1 + R

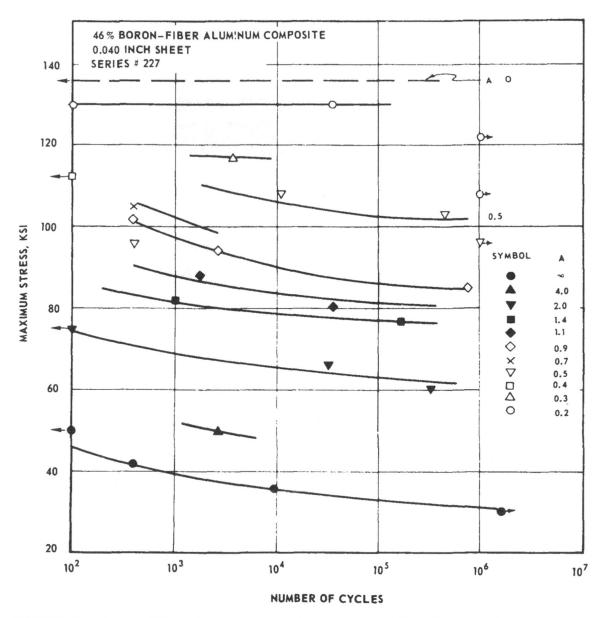

FIGURE 127. Influence of "A" ratio on the measured fatigue properties of 46 vol % boron-aluminum unidirectional composites.

load angle increases the fatigue strength for the unidirectional and ±5° angle-plied composites approaches the fatigue strength of the matrix (16,000 psi at 10^7 cycles). The ratio of fatigue strength to tensile strength is given for each data point. There is no systematic variation of this ratio with filament-load angle. Note the similar angular dependence of the fatigue and tensile strengths (Figure 112). The data for the unidirectional and angle-plied composites are not directly comparable. For the unidirectional composites all the fibers form the given angle with the load axis. For the angle-plied composites the angle cited is the angle that the load axis makes with the axis of symmetry of the composite. This means that a ±5° angle-plied composite tested at a "filament-load" angle of 25° will have half of the filaments at 20° and the other half at 30° to the load axis.

The fatigue strength for the ±45° angle-plied composites tested at 45° to the composite axis of symmetry is about 57% of the 0° unidirectional fatigue strength.[186,189] This observation can be

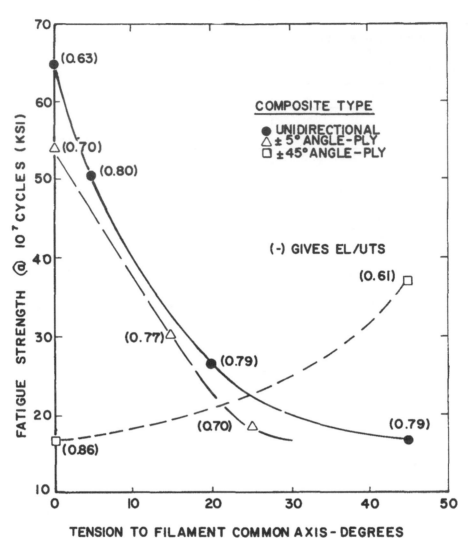

FIGURE 128. Variation of fatigue strength at 10^7 cycles with filament-load angle for various filament configuration of 25 vol % boron/6061 aluminum composites.

rationalized by noting that the number of boron filaments oriented to contribute directly to the composites axial fatigue strength has been reduced by a half. In the case of the ±45° angle-plied composites tested parallel to the axis of symmetry, extensive load transfer through the matrix is required which would account for their relatively poor fatigue strength. The observed fatigue strength is essentially equal to the matrix fatigue strength. A limited number of fatigue tests have been performed at 90° to the filaments for unidirectional B/6061 Al composites.[179] The 90° fatigue strengths for 25 and 50 vol % boron composites with an "R" ratio of 0.1 were found to

be 8,000 psi and 4,000 psi, respectively. These fatigue strengths are poor when compared with the fatigue strength of the matrix, even when the filaments are assumed to make no contribution to composite fatigue strength (i.e., filaments replaced by circular cylindrical voids). The explanation of these low 90° fatigue strengths is probably related to pre-existing or stress-induced defects. These defects may take the form of cracked filaments, poor filament-matrix bonds, poor matrix-matrix bonds, or the presence of cracked and/or brittle reaction products. Another investigation has reported somewhat improved 90° fatigue properties for similar material (46 v/o Borsic/6061 Al) tested under similar conditions (tension-tension low cycle fatigue, R = 0.1).[187] These results are shown in Figure 129.

Fatigue Failures

Composite fatigue failures, in general, have been found to originate at stress concentrations usually located at the filament-matrix interface. The following failure sequence has been reported for unidirectional boron-aluminum composites tested in tension-tension axial fatigue parallel to the filaments.[186,189] (1) Failure is initiated at pre-existing filament breaks, at filament breaks formed on initial application of load, or at surface filament breaks resulting from specimen preparation. (2) The stress concentrations in the matrix resulting from cracked filaments increase in intensity with time as a result of matrix strain hardening. (3) Eventually these stress concentrations become intense enough to initiate cracks at an adjacent filament-matrix interface. Crack initiation will be facilitated by high matrix strain hardening rates and the related higher stress levels. (4) These cracks propagate along the filament-matrix interface until a weak point is found in the filament. The filament breaks and the process is continued. Cracks have been observed to propagate through a series of adjacent filaments without propagating through the intervening matrix. Composite failure takes place when the crack exceeds the critical length established by reinforcement content, matrix properties, and specimen dimensions.

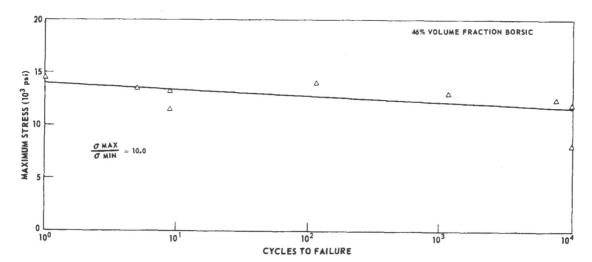

FIGURE 129.   Transverse fatigue curve for 46 vol % Borsic-6061 aluminum unidirectional composite.

Chapter 10

IMPACT

The impact energy for 50 vol % Borsic-aluminum composites as determined from full size Charpy "V" notch specimens is shown in Figure 130.[1 2 3] The "LT" notch-filament configuration not only yields the highest impact energy but also demonstrates an increase in energy absorption capacity with increasing boron content. In comparison, the "TT" and "TL" notch-filament configurations have much lower energy absorption capacity and appear to be insensitive to boron content. In the case of the "LT" notch-filament configuration, the crack front is propagating normal to the filaments with composite failure requiring the fracture of all filaments in the cross section. With this configuration cracks can be deflected parallel to the filaments along the filament-matrix interface increasing the composites energy absorption capacity. The stresses normal to the filament are considerably smaller than is the case for the other two configurations.

In contrast, cracks propagating in the "TT" and "TL" specimens are not required to fracture all

the boron filaments in the cross section. Instead, the high transverse stresses acting on the filament near the crack tip cause the filament to split longitudinally. The stress required to cause longitudinal filament splitting (~30,000 psi) is considerably less than the stress required for the fracture of boron filaments normal to their axis (~450,000 psi). Although the strengths cited above cannot be directly applied to impact energy predictions, they do indicate that the filament will be able to make a larger contribution to the composites energy absorption capacity in the "LT" configuration as compared with the "TT" and "TL" configurations. In addition, longitudinal filament splitting can provide an easy path for crack propagation in the case of the "TT" and "TL" configurations. This conclusion was substantiated when extensive filament splitting was observed at the fracture surface of these specimens. The slight improvement of the "TT" configuration relative to the "TL" configuration may result from a larger matrix contribution to impact energy in the "TT"

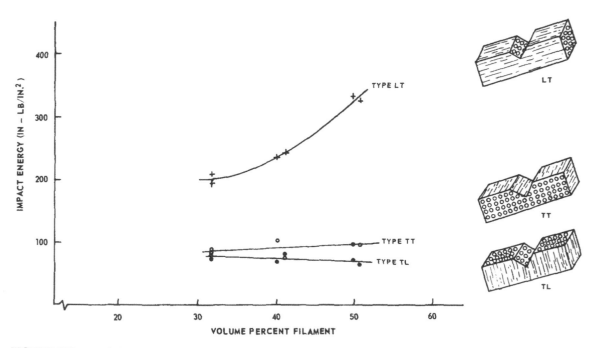

FIGURE 130.   Variation of impact energy with filament content for various notch-filament configurations for 46 vol % Borsic-6061 aluminum unidirectional composites.

case. More extensive plastic deformation is possible around the filaments which run parallel to the crack front in "TT" specimens. In comparison, relatively little plastic deformation can take place in the "TL" specimens without transferring loads into the filaments since the slip bands will necessarily intersect the filaments.[1,2,3] Here again, filament splitting seems to be a limiting condition.

Chapter 11

ELEVATED TEMPERATURE TENSILE STRENGTH

The variation of ultimate tensile strength of unidirectional B/Al composites with test temperature is shown in Figure 131.[36,179,186,190] For comparative purposes, the variation of tensile strength with temperature for 6061 aluminum is also shown. The most important observation is that the composites retain their strength exceptionally well up to about 600°F (315°C). At 600°F (315°C) the composite tensile strengths are still 10 to 30 times higher than the tensile strength of the aluminum alloy matrix. The variation of ultimate tensile strength of cross- and angle-ply B/Al composites is shown in Figure 132.[179,186] Notice that the ±5° angle-ply and 0° to 90° cross-ply composites have the same temperature dependence as the unidirectional composites shown in Figure 131 while the ±30° angle-ply composites reflect the decreasing matrix strength

above 300°F (149°C). The 0° to 90° cross-ply composites are being tested parallel to one half of their filaments so that this temperature dependence would be expected. The ±5° angle-ply composites have the same temperature dependence because they are less sensitive to matrix shear strength and because, for short gage lengths, a majority of the filaments extend from grip-to-grip. In comparison, extensive load transfer through the matrix is required for the ±30° angle-ply composites which is reflected in their stronger temperature dependence.

The temperature dependence of boron filament strength is not presented in Figure 131, but strength reductions of 20 to 40% have been reported from room temperature to 750°F (400°C) for filament with a room temperature tensile strength of approximately 500,000 psi.[24,190,191] The observed composite strengths at 750°F (400°C) are less than rule-of-mixtures predictions, even if a 40% filament strength reduction is assumed. The synergistic influences of residual stresses and matrix constraint are less pronounced at the higher test temperature than at room temperature. As has been discussed, the direct application of filament tensile data to composite strength predictions is oftentimes quite misleading, particularly at elevated temperatures where the chemical reactivity of the filament with the atmosphere or metal matrix introduces an additional complexity.

The variation of transverse tensile strength with test temperature for 25, 37, and 50% B/6061 Al composites is shown in Figure 133.[179] The temperature dependence of the transverse strength of these composites is independent of the reinforcement content. The transverse strength decreases with increases in filament content at all temperatures, which is consistent with the data presented in Figure 109. The orientation dependence for unidirectional 25 vol % B/6061 Al composites at room temperature, 400°F (205°C), and 600°F (315°C) is compared in Figure 134.[186] The variation of fracture surface appearance with test temperature for a unidirectional composite tested at 20° to the filament orientation is shown

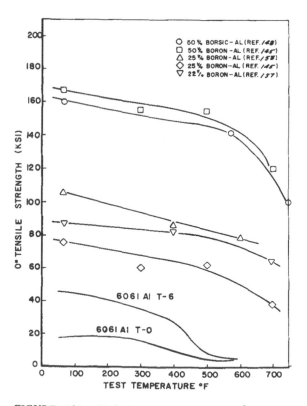

FIGURE 131. Variation of longitudinal (0°) tensile strength with test temperature for unidirectional boron (Borsic)-aluminum composites.

107

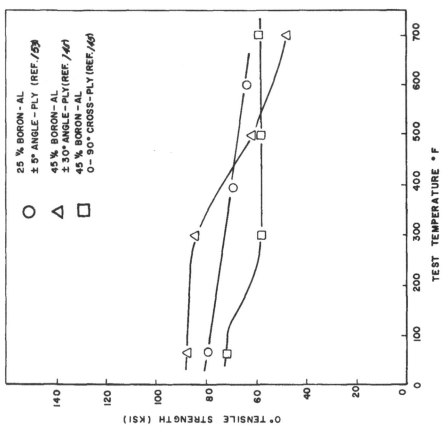

FIGURE 132. Variation of longitudinal (0°) tensile strength with test temperature for various cross- and angle-plied boron-aluminum composites.

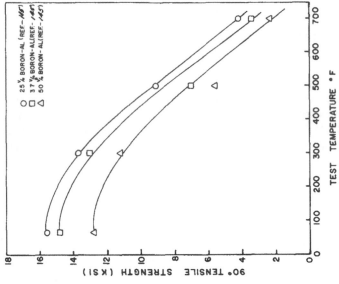

FIGURE 133. Variation of transverse (90°) tensile strength with test temperature for unidirectional boron-aluminum composites.

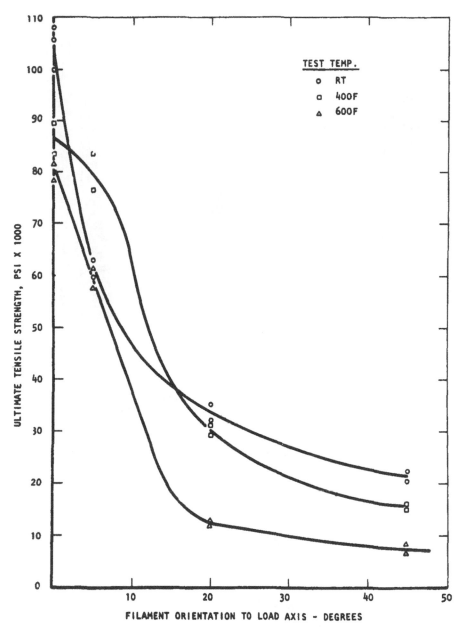

FIGURE 134. Variation of longitudinal (0°) tensile strength with filament-load angle at room temperature, 400°F and 600°F for 25 vol % boron/6061 aluminum composites.

in Figure 135.[186] Note the increasing amount of matrix shear parallel to the filaments. At 600°F (315°C) the fracture surface is parallel to the filaments. The variation of elastic modulus with test temperature and boron content is shown in Figure 136.[179,186] As expected, the longitudinal modulus increases with increased boron content. The maxima at 300° to 400°F is most likely associated with relaxation of residual stresses, both in the matrix and filament. This unexpected maximum was not observed by two other investigators.[173,192] Instead, they found a continuous drop in longitudinal modulus with increasing test temperature. At 600°F (315°C) the room temperature modulus had been reduced 20 to 30%.[173] The transverse modulus was found to be less sensitive to boron content, decreasing with increasing test temperatures. Room temperature tensile strength after elevated temperature exposure has already been discussed.

70°F   400°F   600°F

FIGURE 135. Change of tensile fracture surface appearance with test temperature for unidirectional boron-aluminum composites tested at a filament-load angle of 20°

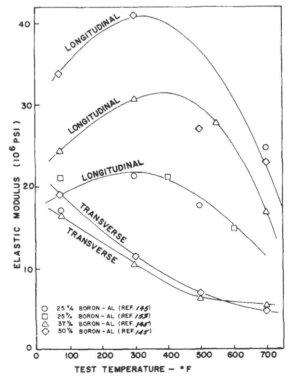

FIGURE 136. Variation of longitudinal (0°) and transverse (90°) elastic modulus with test temperature for unidirectional boron-aluminum composites.

Chapter 12

CREEP AND STRESS RUPTURE

The creep resistance of unidirectionally reinforced metals loaded parallel to the filaments has been found to be outstanding, primarily because of the creep resistance of the filament. Several models have been proposed to explain the creep behavior in fiber reinforced metals; but all conclude that, at least in the direction of the fibers, the properties of the reinforcement determine the creep properties of the composites.[193,194] At 1500°F (815°C) with a 200,000 psi applied stress, the creep strain of boron filament was found to be approximately 1% after 275 hr.[195] At 1000°F (538°C) the minimum creep rate of boron filament under a stress of 264,000 psi was found to be 1×10^{-5} in./in./hr.[196] At 1500°F under 200,000 psi stress the minimum creep rate of boron filament was found to be 4.5×10^{-4} in./in./hr.[195] The minimum creep rates of boron filament coated with a thin ($<0.5\mu$) layer of aluminum and tested at 500°F under a stress of 100,000 psi were found to be about 5×10^{-6} in./in./hr.[190] These properties are superior to the creep properties of tungsten wire tested under similar conditions. At the same fraction of their melting points ($0.45\ T_{mp}$) boron filament required a stress of 220,000 psi while tungsten wire of similar diameter required a stress of 40,000 psi to derive a creep rate of 10^{-3} in./in./hr.[195]

Unlike homogeneous alloys, the creep of B/Al composites has been found to be relatively insensitive to test temperature.[190] For 22 vol % B/Al composites studied from 200° to 600°F (93° to 315°C) under a constant stress of 50,000 psi there was about a tenfold increase in the minimum creep rate ($0.15 \rightarrow 1.8 \times 10^{-5}$ in./in./hr.) and a tenfold increase in creep strain ($0.1\% \rightarrow 1\%$) with an apparent activation energy of 3,000 calories.[190] The corresponding activation energy for the matrix is typically 30 to 35,000 calories. In comparison, the apparent activation energy for creep of a unidirectional 60 vol % B/Al composite tested in three-point bending under a nominal stress of 160,000 psi was about 30,000 calories.[178] This higher activation energy implies that the flexural creep process is controlled by matrix shear.

Creep curves for 25 vol % B/Al composites tested at 400°F (204°C) under varying initial stresses are shown in Figure 137.[186,189] Note that the initial stress level on the unidirectional composites tested parallel to the filaments (0° U-D) was much larger that the initial stress for the other orientations. Creep curves for the 0° unidirectional composites exhibit three regions: the initial region where strain results from instantaneous load application; a region where the matrix relaxes, resulting in additional elastic extension of the filaments; and a secondary creep region where creep is probably controlled by the creep of the filaments. A terminal region of accelerated creep was usually not observed.[186,189] At filament-load angles other than 0°, the creep resistance of the composites becomes dependent upon the creep properties of the matrix as

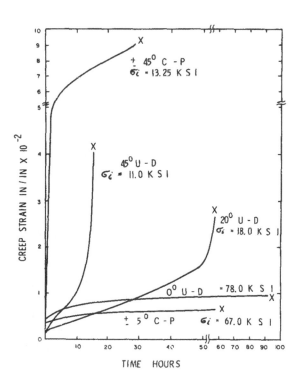

FIGURE 137. Creep curves of some unidirectional (U-D) and cross-plied (C-P) 25 vol % boron/6061 aluminum composites at 400°F.

evidenced by increasing creep strains and minimum creep rates as the filament-load angle increases. In the case of the ±45° cross-ply composite, radiographic analysis after testing revealed that the filaments had rotated toward the load axis assuming a final orientation near the fracture surface of ±28°.[186,189] This "scissoring" action leads to the high observed creep strains. The minimum creep rate for the ±45° cross-ply composites is lower than the minimum creep rate for the 45° unidirectional composites. Examination of the fracture surface for the 45° unidirectional specimen showed that failure took place by a matrix/shear-controlled process parallel to the filaments.[186,189] In comparison, for the ±45° cross-ply composites, extensive shear in the matrix parallel to one set of filaments is impeded by the second set resulting in a lower minimum creep rate. The effect of stress on the minimum creep rate of 25 vol % B/Al composites at 400°F (204°C) and 600°F (315°C) for several filament orientations is shown in Figure 138.[186,189] Notice the stronger temperature dependence for the ±45° cross-ply and 45° unidirectional composites compared with the 0° unidirectional composites. This reflects the requirement for extensive load transfer through the matrix and, consequently, a greater sensitivity to matrix properties in the ±45° and 45° composites.

The stress rupture properties of filamentary reinforced metals have also been found to be excellent, especially when loaded parallel to the filaments. Figure 139 shows the 0° stress rupture behavior of 50 vol % Borsic/Al composites at 300°C and 500°C and shows, for comparison, the stress rupture behavior of titanium 6Al-4V at 500°C.[36,39] The superiority of the composites is obvious. Figure 140 compares the 1,000 hr rupture stress of 50 vol % Borsic/Al composites with that of titanium 6Al-4V.[36] The improved performance of the composites becomes particularly significant for temperatures greater than 400°C where the strength of titanium 6Al-4V drops quite rapidly. A comparison of the stress rupture behavior of 50 vol % Borsic and boron reinforced aluminum composites at 400°C is shown in Figure 141.[36] The greater chemical

stability of Borsic filament is reflected in a substantially lower stress dependence of the rupture time, although for short times the data merge and become nearly equivalent.

Figure 142 shows the stress rupture behavior of 25 vol % B/Al composites with various filament orientations.[186] Here, again, note the superiority of the ±5° angle-ply composites relative to the 5° unidirectional composites. This is also evidenced for ±45° cross-ply and 45° unidirectional composites in Figure 143.[186,189] Notice that as the filament-load angle increases the rupture stress decreases. This effect is summarized in Figure 144, where the 100 hr rupture stress is found to decrease with increases in filament-load angle, much like the filament-load angle dependence of tensile strength (Figure 112) and fatigue strength (Figure 128).[186,189] Although the effect has not been documented in the boron filament reinforced systems, a linear relationship has been found between fiber content and rupture life at constant stress and temperature for tungsten wire reinforced nickel base alloys.[143]

Edge and center initiated filament cracks have been identified on composite fracture surfaces.[36,39] Edge initiated filament breaks result from surface flaws and commonly have a fan-like cross section containing many wedge-shaped filament flakes. The center initiated crack has its origin at the boron-core interface and propagates roughly normal to the filament axis. High residual compressive stresses (70,000 psi) are found at the surface of the filaments with compensating tensile stresses in the interior.[197] These residual stresses will promote center and inhibit edge initiated filament fracture. As the test temperature is increased a larger percentage of edge initiated cracks is noted. This increase may be related to the partial relaxation of residual stresses in the filament. The fracture surface at room temperature for 0° unidirectional composites is roughly normal to the filaments with little filament pullout. As the test temperature is increased, larger amounts of matrix shear and filament pullout are observed. The extent of filament pullout, however, is observed to be less than the simple shear-lag theory would predict.[39]

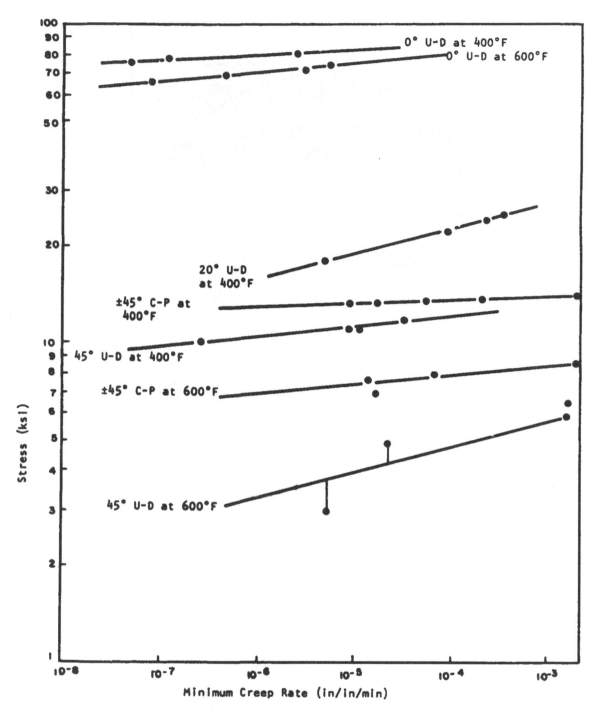

FIGURE 138. Effect of stress on minimum creep rate.

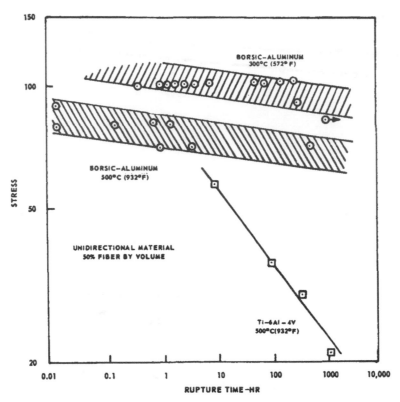

FIGURE 139.   Stress-rupture properties of Borsic-aluminum composites.

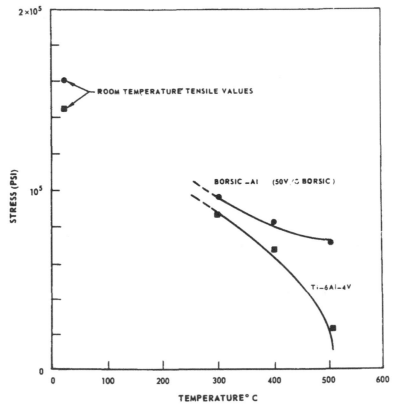

FIGURE 140.   Stress for 1000-hr rupture life as a function of temperature.

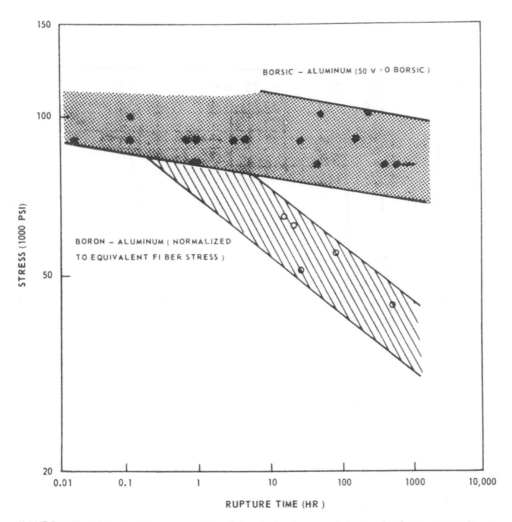

FIGURE 141. Stress-rupture properties of Borsic-aluminum and boron-aluminum composites at 400°C.

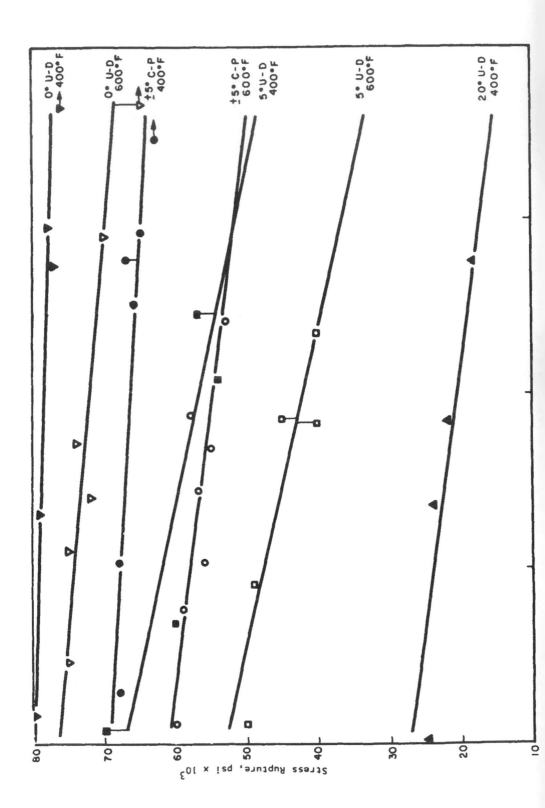

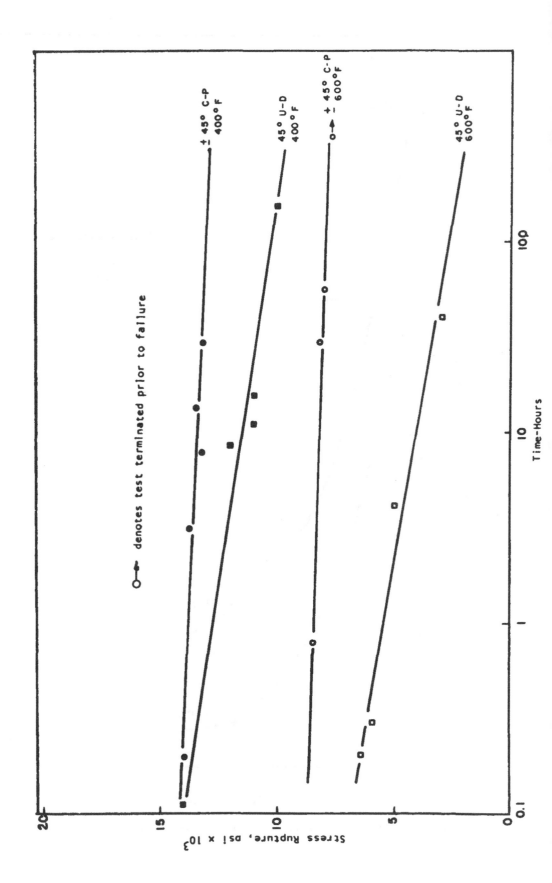

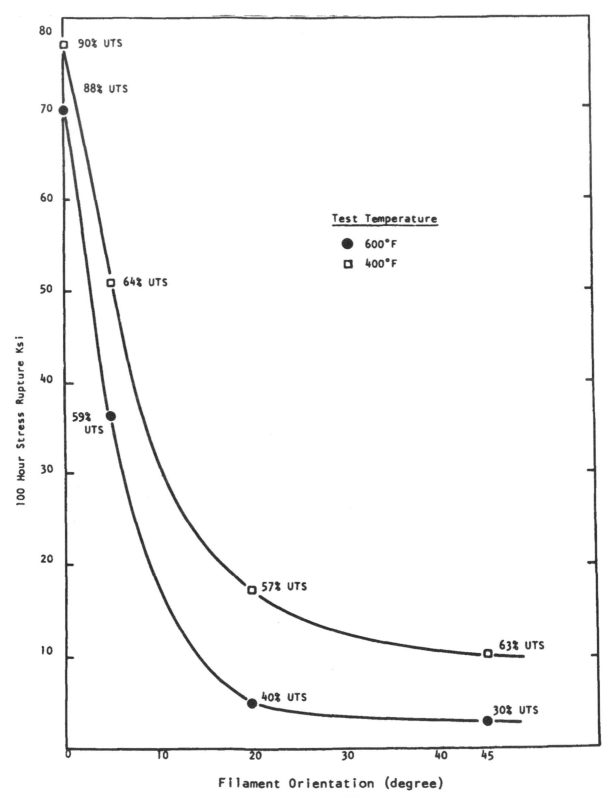

FIGURE 144.   100-hr stress-rupture as a function of filament orientation of 25 vol % boron/6061 aluminum composites.

Chapter 13

EUTECTIC COMPOSITES

Introduction

The composite structures that we have been considering thus far have been filamentary reinforced metals where independently generated reinforcements have been consolidated in a metal matrix. In this section we will consider composites made from the unidirectional solidification of eutectic alloys. Unidirectional solidification is an exception to the methods which have been presented. In this process the reinforcing phase is grown in situ during controlled solidification from the melt.[198] Fibrous or lamellar structures are produced depending on the system, the

composition of the system, and growth conditions such as the rate of solidification. In Figure 145[199] the two types of morphology are shown. The Al-Al$_3$Ni consists of rods or fibers of Al$_3$Ni in the Al matrix. The Al$_3$Ni shows as the dark, discontinuous, but aligned phase at approximately 11 vol %. In the Al-CuAl$_2$ system the dark phase is the CuAl$_2$ lamellar structure which comprises 45 vol % of the eutectic composite.

Unidirectional solidified composites have the advantage that the reinforcing phase is grown on low energy planes in equilibrium with the metal matrix.[199] This provides high temperature

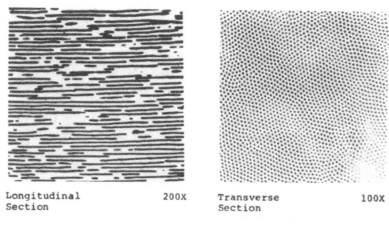

Longitudinal   200X
Section

Transverse   100X
Section

A. Al-Al$_3$Ni Exhibiting rod like whiskers.

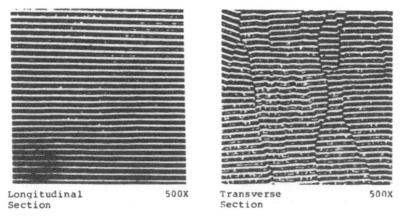

Longitudinal   500X
Section

Transverse   500X
Section

B. Al-CuAl$_2$ exhibiting lamellar structure.

FIGURE 145.   Eutectic composite microstructures.

stability for the phases because there is little driving force for reinforcement matrix reactions. The microstructure and properties of Al-Al$_3$Ni have been found to remain stable after 500 hr at 80% of the eutectic temperature.[200] Although the Al-Al$_3$Ni melting point is only 640°C the high temperature strength exceeds that of a precipitation hardened aluminum alloy, 7075-T6.

A turbine blade shape[199] has been directionally solidified in the Ni-NiMo system. This eutectic has a lamellar type microstructure and compares favorably with nickel superalloys up to 1800°F (981°C) and is stronger at 2000°F (1093°C). This latter work points to a prime emphasis on the potential of eutectic composites to develop new and improved high temperature materials for gas turbine engine stator vanes and rotor blades. Target temperatures for this type application greater than 2000°F (1093°C) are common and could not be met by most filamentary composites because of severe matrix filament reactions and consequent property degradation.

In binary eutectic systems the amount of reinforcing phase is dictated by the phase diagrams. In practical systems of interest, many have an insufficient amount of reinforcing phase to make them competitive with conventional alloys. Research in recent years has led to the development of more complex eutectic compositions with attendant flexibility in compositional variations. Foreign element additions to surprisingly large concentrations can sometimes be made while still maintaining the basic composite structure as a means of improving specific properties such as oxidation resistance and phase stability. Ternary alloys may be solidified along the eutectic trough. One example of a more practical system is Ni$_3$Al-Ni$_3$Nb.[201] This eutectic composite and several others of practical import will be considered in more detail in this section.

Eutectic composites also have elicited considerable interest for nonstructural applications because of their anisotropic thermoelectric, magnetic, and optical properties. One particular promising controlled (aligned) eutectic system is that of InSb-NiSb. This composite contains NiSb needles in an InSb matrix. This material has been proposed for a number of electrical devices.[202] The properties of this system have been reviewed in some detail recently[203] and will be considered later in this section.

Other approaches to the development of metallic materials for high temperature use closely parallel work on directionally solidified eutectics. This includes the related directional solidification of nickel-base superalloys to achieve a columnar microstructure. Strengthening in this case is primarily associated with the presence of precipitates. These precipitates do not remain in a stable high strength structure at temperatures close to the melting point as do the eutectic composites, but do represent a significant advance in metallurgical structures. From a practical standpoint the directional solidification techniques developed for these alloys provide the background for the scaling up of the development of directionally solidified eutectic composites. Another approach has been through dispersion strengthened alloys such as the use of thoria additions to nickel, nickel-chrome, and cobalt-chrome alloys. These alloys possess higher temperature strengths but at a loss of ductility Additions of greater than a few vol % of the oxide phase produces severe embrittlement. Therefore, although these are composite materials in the classical sense, they have not been considered in this book, where emphasis has been placed on continuous filament reinforcement. Even short discontinuous filaments are much longer than the dispersed phase materials. It should also be noted that the fibrous second phase in a eutectic composite like Ni-Al$_3$Ni is like a very long whisker in morphology.

Growth Techniques

The laboratory investigation of directionally solidified eutectics uses techniques which are not always suitable for larger scale commercial processes. Most investigators have used either induction or electron beam heating or a modified Bridgeman technique to form eutectic composites. A few investigators have used chills, insulation, and special heating to generate larger specimens with "state of the art" casting techniques. Unidirectional solidification using inductive heating is usually conducted vertically using a precast and homogenized rod as the starting material. Casting is accomplished either in a vacuum or in an inert gas atmosphere to avoid surface oxidation, dissolution of gases, and voids in the solidified material. The rod is contained in a suitable crucible material (graphite, boron nitride, etc.) which is contained in a closed quartz tube to permit atmospheric control. The induction coil is moved up the quartz tube at a fixed rate which can then be equated to the rate of solidification. Thermal gradients are increased by chilling the crucible just below the induction coil. Critical temperatures are measured either by pyrometry or specially placed traveling thermocouples, although analytical techniques may be required to generate the desired data from these measurements. Artificial convection in the melt can be introduced by the rotation of the upper portion of the rod, although this rotation has, on occasion, been found to influence the macrostructure of the solidified specimen.

The electron beam technique is particularly well suited to systems where the molten metals tend to react with the crucible material (i.e., titanium base eutectics). The size and geometry of specimens produced by this technique are somewhat restricted by the capabilities of the electron beam process. The electron beam process is operated under a vacuum of 10^{-5} or 10^{-6} mm Hg. The size of the molten zone can be controlled by variation of the applied potential, filament current, and/or the positioning of the focusing plates. Precast and homogenized rods are supported vertically in a vacuum chamber while the filament is passed up along the length of the rod which is supported at both ends. Passage of the molten zone down the rod generates a less stable condition. The rate at which the filament is moved up the rod can be equated to the rate of solidification. Temperatures are reliably measured by pyrometry. A water cooled chill is moved up the rod just behind the filament to induce larger thermal gradients at the solid-liquid interface. Artificial convection can be promoted by rotation of the upper portion of the rod. Experimental control of crystallographic orientations can be accomplished using a single crystal "seed" at the initial (lower) section of the starting rod.

The modified Bridgeman technique basically involves the vertical solidification of eutectics by slowly raising the furnace surrounding the specimen, thereby allowing the controlled solidification of the sample. (Although it might seem easier to lower the specimen than raise the furnace, the increased stability associated with holding the specimen fixed often yields a superior result.) The entire specimen is initially molten and is contained in a suitable crucible. The specimen and crucible are characteristically contained in a closed quartz tube to permit atmospheric control. Larger thermal gradients are induced with a water cooled chill which is usually an integral part of the furnace. The chill is located just below and very close to the hot zone of the furnace. Temperatures are measured with prepositioned thermocouples. The rate of solidification can be equated to the rate at which the furnace is raised.

Commercial casting processes are modified to generate unidirectional heat flow and large thermal gradients. Hot tops and water cooled chills are used to control the solidification rate and the thermal gradients. Insulated molds are used to promote unidirectional heat flow. The combination of heating the top, cooling the bottom, insulating the sides, and controlling the degree of superheat of the molten metal has been found to offer sufficient control to yield the characteristic directionally solidified eutectic microstructure. Obviously the degree of control available is less than with the previously cited methods. The microstructural outcome, therefore, is determined by the sensitivity of individual material systems to the magnitude and control of such important parameters as the thermal gradient in the liquid at the solid-liquid interface and the local solidification rate.

Growth Conditions

Factors which influence the solidification process and the resultant microstructure of eutectic composites include the previously mentioned thermal gradient in the liquid at the solid-liquid interface and the local solidification rate. The composition and purity of the alloys and the degree of convection in the molten zone are important variables. Mollard and Flemings[204,205] have reviewed the growth of composites from the melt. They have concluded that plane front solidification to achieve regular spacing and alignment in eutectic microstructures is favored by

1. low growth rate
2. high thermal gradients
3. absence of convection in melt.

Their theoretical treatment[204] was followed by a demonstration of optimizing growth conditions in the Pb-Sn system.[205] The alloys were prepared over a wide range of concentrations. Structures of alloys of near eutectic (26 at. % Pb) composition, solidified with a plane front, exhibited a lamellar structure. Alloys farthest from the eutectic (12 at.

% Pb) exhibited a rod-like morphology. Structures of ingots as a function of the temperature gradient/growth rate ratio were determined. The spacing between phases increased with decreasing growth rate, as has been generally observed with eutectic composites.

The solidification rate is straightforwardly controlled when the induction, electron beam, and Bridgeman techniques are employed. Even the modified commercial casting process offers considerable control over the macroscopic or average rate of solidification. Control of solidification rate becomes more difficult with increased section size and variable section geometry.

With the extremely high solidification rates which are only achievable with small specimens, the minor phase is found to be dispersed in the matrix in the form of very fine uniformly dispersed particles. At the other extreme, very low rates of solidification result in either a very coarse eutectic or a spherodized microstructure. Between these microstructural extremes dendritic, cellular, plate-like, or rod-like microstructures may be derived, depending on the material system and the solidification conditions.

The tensile strength in the direction of the reinforcement phase in the $Al-Al_3Ni$ system increases with increasing solidification rates. This was observed over a wide range of solidification rates from 2 to 38 cm/hr.[206] The transverse tensile strength is higher at much lower growth rates, however. At slower rates of solidification the finely dispersed needles of Al_3Ni were found to change to coarser, flatter, blade-like structures which gave a maximum transverse strength at growth rates as low as 0.35 cm/hr.[207] Below this rate some irregularities in the structure occurred which were attributed to a "degenerate microstructure" previously observed for the $Al-CuAl_2$ system by Chadwick.[208] While the decrease in transverse strength is not great, it indicates the need for optimizing growth conditions with concern for more than the uniaxial tensile strength as the sole criterion of results.

The rate of solidification, of course, is determined by the rate of heat loss from the system. If no additional heat is provided to the system then the solidification rate and the thermal gradient at the solid-liquid interface are directly related. The more rapidly heat is removed from the system the higher the solidification rate. If heat is added at the same rate that heat is being lost, the solidification rate will be zero but there will still be a substantial thermal gradient in the liquid at the solid-liquid interface. Finally, by adding heat at a lower rate than heat is being withdrawn one has some control over both the solidification rate (by the difference between heat input and heat loss) and the thermal gradient (by the difference in temperature). For large samples the alloy's thermal conductivity becomes a limiting factor.

High thermal gradients in the melt adjacent to the solid-liquid interface promote the stability of a planar interface. In effect, local interfacial perturbations initiated by some local instability are suppressed by the thermal barrier they face. A stable planar interface is a prerequisite to formation of the desired eutectic microstructure. For analytical purposes the solidification rate by itself is often less valuable than such multivariable factors as G/R (thermal gradient/solidification rate) in determining the microstructural result of directional solidification. One can conclude with some reservations about optimized conditions that low growth rates and high thermal gradients in the melt promote the stable, ordered, and aligned growth of the eutectic microstructure during directional solidification.

The reservations technically include the low solidification rates producing lower energy, more stable, coarse microstructures with lower strengths and creep resistance. On a practical basis they must also include the increased cost of control of production methods for slow growth rates. The finer microstructures from rapid solidification of the $Al-Al_3Ni$ eutectic coarsen on heat treatment at temperatures of 500° to 600°C.[200] The change is attributed to a reduction in the interfacial area per unit volume in the transverse direction since the Al_3Ni whiskers (rods) do not change in the longitudinal, axial direction. The uniaxial strength is also relatively constant. The same type of coarsening of the lamellar $CuAl_2$ in the $Cu-CuAl_2$ eutectic is observed on heat treatment at 500° to 540°C, resulting in a reduction in interfacial area.[209] The effects of heat treatment and cold rolling on eutectic composites have been reviewed by Salkind.[210]

The potential for solidification of various binary eutectic alloys to produce ingots of controlled eutectic composite structures has been reviewed by Kraft.[198] This reference provides an excellent background to the broad possibilities for an extensive variety of alloy types which form suitable structures. Hunt and Jackson[211] have also reviewed the binary eutectics, giving examples of the irregular and complex structures that are observed under varying growth conditions. In more complex eutectic compositions Yue[212] has shown that experimental analysis of compositions from the zone melting technique can be used to determine eutectic compositions. This has been demonstrated for a quaternary system, Al-Mg-Zn-Si, in which the eutectic composition is unknown and in the complex system, Al-Mg-Zn-Cu-Sn-Pb, in which the eutectic composition cannot possibly be determined by conventional analysis of phase diagram data. Theoretical arguments of Mollard and Flemings[204] and practical demonstrations in the Pb-Sn system[205] as previously described have shown that it is possible to maintain a plane front in alloys far from the eutectic composition, at reasonable growth rates and thermal gradients. One of the requirements is the essential absence of convection. Verhoeven and Homer[213] have pointed out that this condition is a severe restriction in some systems, but have presented theoretical arguments on the growth of aligned off-eutectic composites from stirred melts. Convective mixing is proposed to enable higher temperature gradients to be maintained in order to stabilize the planar eutectic interface at the off-eutectic compositions. The application of off-eutectic growth techniques to a system such as Al-Al$_3$Ni is quite interesting since the normal eutectic concentration of the Al$_3$Ni reinforcing phase is only 11 vol %.

While it is beyond our scope here to consider the specific conditions which lead to various reinforcement morphologies, some mention of the available literature should be made. Jackson and Hunt[214] have reviewed the conditions for lamellar and rod (fibrous) eutectic growth. They have considered the conditions for stability of rod and lamellar structures and the mechanisms by which structural faults and instability arise. The formation of faults in eutectic alloys has recently been considered by Cline.[215] These are largely mathematical treatments, but numerous observations of the growth structures and spacing changes during fault formation have been made. This includes observations on the nucleation and growth of Pb-Sn eutectics by Hopkins and Kraft,[216] by Cline,[217] and by Jackson.[218] Examples of other investigations of this type include that of Bertorello and Biloni[219] on the cellular substructure of Al-Cu eutectics, and the analysis of the dislocation arrays at interfaces in a NiAl-Cr eutectic by Walter, Cline, and Koch.[220]

The transition between various eutectic structures, particularly between the basic lamellar to fibrous structures, has been the subject of special interest. Structural changes are of great import to mechanical properties and need this attention in any system to be considered on a practical basis. Jaffrey and Chadwick[221,222] have studied the nucleation, growth stability, thermal stability, orientation relationships, and the lamellar to fibrous transition of Sn-Zn and Al-Al$_3$Ni eutectic composites in considerable detail. The structures, faults, and "rod-plate" transition in Ni-W, NiAl-Cr, and NiAl-Cr(Mo) eutectic systems were also recently studied.[223]

Yue and Crossman[224] used a controlled eutectic solidification technique to produce a Ti-Ti$_5$Ge$_3$ eutectic composite with excellent alignment at concentrations of the reinforcing phase of Ti$_5$Ge$_3$ considerably in excess of the amount present at the eutectic composition (34 vol % vs. 19.5 vol %). The structure of unidirectionally solidified Al-AlSb eutectic composites has been studied over a wide range of growth rates from 0.6 to 49 cm/hr.[225] A broken lamellar structure was produced that appeared to consist of perforated lamella and branched ribbons with some rods present at lower solidification rates. The shape instabilities of eutectic composites have been calculated for a rod composite structure at elevated temperatures by Cline.[226] He concludes that two-dimensional coarsening occurs more rapidly than spheroidization in faceted rod structures and in composites containing more than 20 vol % rods. Fault migration dominates the process at the beginning of coarsening. Some fascinating examples of changes in interface dislocations between phases of NiAl-Cr and NiAl-Cr(Mo) eutectics after elevated temperature treatment and quenching or slow cooling to room temperature have been obtained.[227] Lattice parameter measurements of lattice mismatch by x-ray methods were found to be in excellent agreement with the mismatch calculated from observed spacings of the dislocations.

Potential Systems

Many types of materials combinations can be employed to develop controlled eutectic structures, and thus eutectic composites. The emphasis in this discussion is on metallic structures, but eutectic structures are often found with other materials such as oxides, halides, and organic compounds. Some of these compounds have been listed by Galasso[24] in addition to a more complete listing of metallic and intermetallic eutectics than will be considered here. In metals, as pointed out by Kraft,[198] eutectic reactions can be expected and occur in a majority of alloy systems. Examples of eutectic alloys in commerce are those of Al-Si and cast iron. It is only necessary to employ the appropriate conditions of solidification to develop a high degree of orientation of the eutectic structures and thus produce eutectic composites.

Some confusion is found in the literature on how to list the components of a composite. The order we have followed in general is to give the reinforcing agent first, or B/Al, SiC/Ti, and Al_2O_3/Ni. If any convention is followed on eutectic composites, it is one that tends to be just the opposite of the above practice. One finds the matrix listed in first, or Ni-Al_3Ni, Al-AlSb, Ta-Ta_2C, Ti-Ti_5Si_3, and Al-$CuAl_2$. For a lamellar eutectic such as Al-$CuAl_2$ where the two phases are present in a volume ratio approaching one, the distinction would have to be made based on which phase is the reinforcing phase. For a eutectic composite such as InSb-NiSb, the distinction must be based on the electromagnetic and optical effects peculiar to adding the NiSb to the InSb since "reinforcement" as such is not of primary interest.

It is instructive at this point to consider several specific eutectic composite systems which appear to have interesting properties at elevated temperatures. One of these is based on the intermetallic phase, Ni_3Al (γ').[201] This phase is encountered in nickel-base superalloys which are strengthened by precipitates of the γ' type. This compound has, therefore, received considerable study which provides valuable background data for understanding the behavior of a Ni_3Al containing composite. Single crystal Ni_3Al is ductile at room temperature.[228] The critical resolved shear stress reaches a maximum at $1500°F$ ($815°C$), and other elements in solid solution strengthen the Ni_3Al phase.[229] Eutectic systems recently studied have been based on adding another element to this system, for example, Ni-Al-Nb, Ni-Al-Zr, and Ni-Al-Ti.[230] The phases then consist of Ni_3Al with Ni_3Nb, Ni_7Zr_2, and Ni_2TiAl plus Ni_3Ti, respectively. The best characterized of these systems has been that of Ni_3Al-Ni_3Nb, which develops promising properties for elevated temperature strengthening of a reasonably ductile matrix.[201,230]

The properties of Ni_3Al and Ni_3Nb have been the subject of several other investigations[231,232] and the high temperature properties of Ni_3Al-Ni_3Nb alloys have been studied.[233] The phase diagram of the Ni_3Al-Ni_3Nb system is shown in Figure 146.[234] The eutectic composition was found at 70 wt % Ni_3Nb with a eutectic temperature of $1280°C$ ($2336°F$). The solid solubility of Ni_3Nb in the γ' extends to 40 wt % (8.5 at. % Nb) and the solubility of Ni_3Al in δ is less than 4 wt % (1.3 at. % Al). The later work of Thompson and George[201] places the eutectic composition at 68 wt % Ni_3Nb. This corresponds to a composition of 23.1 wt % Nb, 4.4 wt % Al, and the balance of Ni. An optimum growth rate of 2 cm/hr was selected. The microstructure of the eutectic composite is shown in Figure 147. The lamellar morphology has a volume ratio for the phases of approximately one.

The practical effect of the variation in properties of the eutectic composites with growth rates is seen in Figure 148. Over a large temperature range the tensile strength is higher for the composite grown at 2 cm/hr compared to the composite grown at 0.5 cm/hr. The magnitude of the difference remains the same over the temperature range from R. T. to $2000°F$ ($1093°C$), and is, therefore, more significant at the higher temperatures.[201] The interlamellar spacing for the 2 cm/hr growth rate was about 3.6 μm. For 0.5 cm/hr the spacing was not reported, but at 0.3 cm/hr the spacing had increased to about 9.6 μm.

Another important approach to eutectic composite structures for elevated temperature applications is based on the reinforcing of a ductile metal matrix with a refractory metal carbide. The carbides are attractive because of their high temperature stability, high elastic moduli, low densities, and their ready incorporation into alloys by conventional melt techniques. Systems which have been studied such as Ta-Ta_2C[235] often appear deceptively simple. Actually, the phase equilibria are often quite complex, as recent data have shown.[236] This can lead to considerable problems

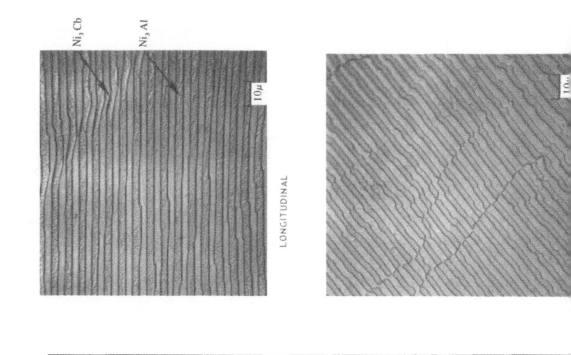

Ni₃Cb

Ni₃Al

LONGITUDINAL

10μ

10μ

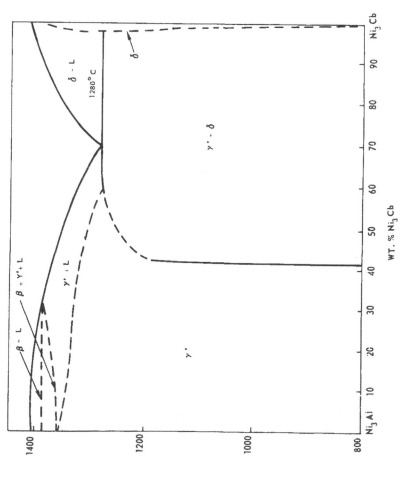

FIGURE 146.   Equilibrium diagram of Ni₃ Al-Ni₃ Cb system.

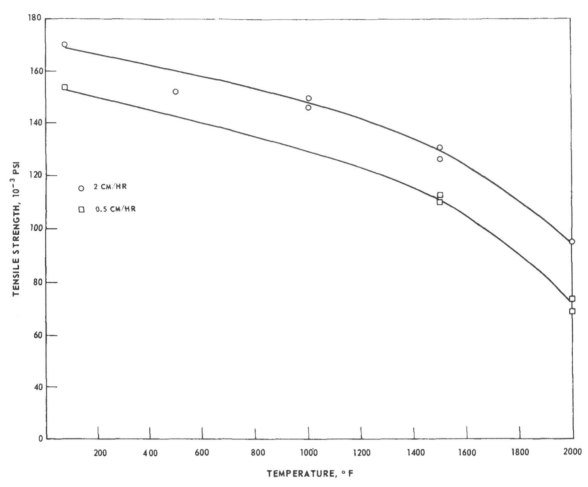

FIGURE 148. Temperature dependence of the tensile strength of Ni$_3$Al-Ni$_3$Cb eutectic unidirectionally solidified at 0.5 and 2 cm/hr.

with nonstoichiometry and phase instability, which are usually conveniently ignored in eutectic composite investigations.

Several fairly attractive systems have been delineated from pseudobinary joins of cobalt and nickel with various refractory monocarbides. Among the systems studied have been Ni-TiC,[237] Co-NbC,[238] and Co-TaC.[239] A general pseudobinary diagram has been utilized by Lemkey and Thompson[240] in their study of nickel and cobalt eutectic composites with refractory carbide reinforcements, as shown in Figure 149. There are a number of limitations to the use of such a diagram, including existence of nonstoichiometry in the carbide phase, preferential solution of the metal from the metal carbide in the cobalt or nickel phase, and in some cases the lack of a single monocarbide phase of the metal carbide in question. Nevertheless, it provides a simple basis

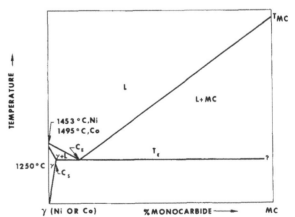

FIGURE 149. Phase diagram for metal-monocarbide system.

for summarizing the possibilities of the metal-metal carbide eutectic composites with metals like cobalt and nickel, where the carbides are stable at

TABLE 18

Eutectic Systems of Cobalt and Nickel with Refractory Monocarbides

SYSTEM	C_S (1250°C)	C_E (Wt %)	VOLUME Carbide (%)	T_E (°C)
Co-NbC	5	11	12	1365
Co-TiC	1	10	16	1360
Co-TaC	–	14	10	1402
Ni-HfC	–	30	15-28	1260
Ni-NbC	3	10	11	1330
Ni-TiC	1.6	4.3	7.5	1307

elevated temperatures. Pertinent data on several of the cobalt and nickel systems are given in Table 18.[240] The Ni-NbC system has been examined in some detail with encouraging results.[240] The reinforcing carbide phase in these composites is fibrous or whisker-like in morphology.

A more complex system of this type is that of (Co,Cr)-$(Cr,Co)_7C_3$ eutectic alloy.[241] The matrix is a Co-28.5 wt % Cr solid solution which has up to 30 vol % of a fibrous chromium carbide dispersion containing a smaller amount of cobalt (approximately 15 wt %). This system also has interesting properties, particularly in the direction of the reinforcing phase.

A brief listing of eutectic systems in metals is given in Table 19.

Mechanical Properties

In many respects eutectic composites constitute laboratory curiosities, with much effort devoted to arriving at new compositions or improving our understanding of perturbations in the existing microstructures. The concomitant effort on characterization of mechanical behavior has been evident on only a few systems, as will be covered here. For many new systems there are little or no property data available. Given the anisotropy of any composite structure plus the variations in rod and lamellar geometries and spacings observed in all systems thus far, the mechanical property characterization becomes a difficult task. It is further complicated by the very brittle character of many of the eutectic phases such as the intermetallic compounds and the metal carbides utilized in the eutectic structures.

The mechanical properties of the Ni_3Al-Ni_3Nb system have been studied in some detail.[201,230]

A few properties have also been determined for the related Ni_3Al eutectics with Ni_7Zr_2, and with Ni_2TiAl plus Ni_3Ti. These systems are not nearly as attractive in properties compared to Ni_3Al-Ni_3Nb.[230] The strength of this eutectic composite in the reinforcement direction is given in Table 20.[201] These unidirectional strengths are quite encouraging. In Figure 150, they have been compared to the strength of an advanced cast nickel-base superalloy called B-1900.[242] The eutectic is of interest mainly at elevated temperatures above about 1800°F (981°C) as seen. Above 2000°F (1093°C) there is a rapid drop in strength. Nevertheless, the strength at 2200°F (1204°C) is nearly 45,000 psi. Thus, the eutectic composite retains considerable strength at temperatures close to the eutectic temperature of 1280°C. The microhardness of the Ni_3Nb, δ phase, was 620 kg/mm^2, of the $Ni_3(Al, Nb)$, γ' phase, 580 kg/mm^2. A single phase casting of Ni_3Al gave a microhardness of 300 kg/mm^2, so the solid solution hardening on eutectic composite formation was substantial. The elastic modulus measured along the growth axis was 35×10^6 psi. The 0.2% offset compressive yield strength was 200,000 psi, tested in the direction of phase alignment. Failure in compression occurred by buckling and interfacial cleavage. Creep rupture data were better in the aligned eutectic than for conventionally cast Ni_3Al-Ni_3Nb bi-phase alloy mixtures.[230] Impact strengths for unnotched Izod and Charpy specimens normal to the reinforcement axis were 3.0 and 12.0 ft-lb, respectively, at room temperature. The impact strength goes up sharply with temperature, exhibiting a maximum of greater than 30.0 ft-lb at 1800°F (981°C). Failure was observed by interlamellar splitting, particularly with increasing temperature. More measurements of off-axis properties which are known to be orientation sensitive are needed to properly characterize systems such as this one. One drawback not mentioned previously is that of fairly high densities, the eutectic in this case calculated at 8.4 g/cm^2 and measured at 8.5 g/cm^2.[201]

The mechanical response of the Ni-Ni_3Nb composite has been investigated under static and dynamic loading conditions.[243,244] Static tests established that deformation of the Ni_3Nb phase occurred primarily by deformation twinning.[243] Tensile fracture initiated at and propagated along specific twin boundaries. Grain boundary fracture was observed occasionally under compressive

TABLE 19

A. Eutectic Composite Systems

Elements	Phases	Composition (w/o or v/o)	Eutectic temp., °C	Structure
Ag-Al	Al-Ag_3Al_2	29.5 w/o Al	566	lamellar
Ag-Cu	Ag-Cu	74 v/o Ag	799	Cu Rods
Ag-Sb	Sb-Ag_3Sb	56 w/o Ag	484	lamellar
Al-B	Al-AlB_2	0.02 w/o B	660	AlB_2 Platelets
Al-Cu	Al-$CuAl_2$	45 v/o $CuAl_2$	548	lamellar
Al-Fe	Al-$FeAl_3$	3 v/o $FeAl_3$	655	$FeAl_3$ Platelets
Al-Ni	Al-Al_3Ni	11 v/o Al_3Ni	641	Al_3Ni rods
Al-Si	Al-Si	11 v/o Si	577	Si rods
Au-Ge	Au-Ge	31 v/o Ge	356	Ge platelets
Au-Pb	Pb-$AuPb_2$	49 v/o $AuPb_2$	215	lamellar
B-Fe	Fe-Fe_2B	53 v/o Fe_2B	1149	lamellar
B-Ni	Ni-Ni_3B	65 v/o Ni_3B	1140	lamellar
Be-Ni	Ni-NiBe	38 v/o NiBe	1157	lamellar
Bi-Te	Te-Bi_2Te_3	27 v/o Bi_2Te_3	413	Bi_2Te_3 platelets
C-Cr	Cr-$Cr_{23}C_6$	3.5 w/o C	1500	$Cr_{23}C_6$ rods
C-Fe	Fe-Fe_3C	59 v/o Fe_3C	1147	lamellar
C-Mo	Mo-Mo_2C	1.8 w/o C	2200	Mo_2C rods
C-Nb	Nb-Nb_2C	31 v/o Nb_2C	2335	Nb_2C rods
C-Ta	Ta-Ta_2C	30 v/o Ta_2C	2800	Ta_2C rods
Cd-Sb	Cd-CdSb	19 v/o CdSb	290	CbSb rods
Co-Mo	Co-Mo_6Co_7	33 v/o Mo_6Co_7	1340	lamellar
Co-Nb	Co-Co_3Nb	39 v/o Co_3Nb	1235	lamellar
Cr-Ni	NiCr-Cr	23 v/o Cr	1345	Cr rods
Cu-Mg	Mg-Mg_2Cu	30.7 w/o Cu	485	lamellar
Fe-S	FeS-Fe	9 v/o Fe	988	Fe rods
In-Sb	InSb-Sb	69 w/o Sb	530	Sb rods
Mg-Ni	Mg-Mg_2Ni	23.5 w/o Ni	507	Mg_2Ni rods
Nb-Ni	Ni-Ni_3Nb	26 v/o N_3Nb	1270	lamellar
Ni-Sn	Ni-Ni_3Sn	38 v/o Ni	1130	Ni platelets
Ni-Ti	Ti-Ni_3Ti	29 v/o Ni_3Ti	1287	Ni_3Ti platelets
Pb-Sn	Sn-Pb	71 v/o Sn	183	lamellar

B. Complex Eutectic Systems

Phases	Composition (w/o or v/o)	Eutectic temp., °C	Structure
CrAs-GaAs	35.4 w/o CrAs	—	lamellar
CrAs-InAs	1.7 w/o CrAs	937	CrAs rods
FeSb-InSb	0.7 w/o FeSb	520	FeSb rods
MnSb-InSb	6.5 w/o MnSb	510	MnSb rods
Ni_3Al-Ni_3Cb	equi-volumes	1280	lamellar
NiSb-InSb	1.8 w/o NiSb	517	NiSb rods
CrSb-InSb	0.6 w/o CrSb	516	CrSb rods

TABLE 20

Tensile Properties of Ni₃ Al-Ni₃ Cb Eutectic Alloy

Unidirectionally Solidified at 2 cm/hr

Test temp., °C	Ultimate tensile strength, psi	Strain at fracture %	Reduction in area %
24	169,800	0.56	—
24	170,100	0.60	—
260	152,600	3.2	1.3
260	152,000	2.7	1.3
538	150,000	8.1	4.5
538	146,000	8.2	7.1
815	126,000	8.4	17.1
815	131,000	5.8	—
981	119,900	1.4	—
1093	94,500	10.0	—
1093	115,300	3.0	16.4
1204	45,200	18.7	7.1
1204	42,100	9.2	7.1

loading. Substantial ductility was noted in this system and attributed to the deformation twinning and subsequent twin boundary fracture of the Ni₃Nb lamellae in the Ni-Ni₃Nb system. A summary of observed properties is given in Table 21.[243] Excellent fatigue resistance was found in notched specimens with an endurance limit of 55% of the 108.7 ksi smooth bar tensile strength. Under high stress-low cycle fatigue, the fatigue resistance of the Ni-Ni₃Nb eutectic composite was controlled by twinning and subsequent twin boundary cracking of the Ni₃Nb phase. Low stress-high cycle fatigue shows crack growth in the nickel matrix by a stage I, crack propagation mechanism. Thus the low stress fatigue life of the Ni-Ni₃Nb system is controlled by the nickel matrix. The Ni-Ni₃Nb composite displays fatigue behavior similar to the Al-Al₃Ni eutectic composite despite the obvious difference in composite structures.[244]

The Ni-Al₃Ni system has been the subject of numerous studies, a few of which will be considered here. The properties have been previously reviewed.[235] The Charpy impact behavior was studied as a function of orientation and test temperature from 400°C to -196°C.[245] The impact strength remained relatively constant at all temperatures. As expected, the impact behavior was quite anisotropic with the lowest impact energy coming from a transverse orientation with the notch normal to the whisker axis. The highest impact energy came from the transverse orientation having the notch parallel to the whisker axis. Composites with blade-like whiskers had higher

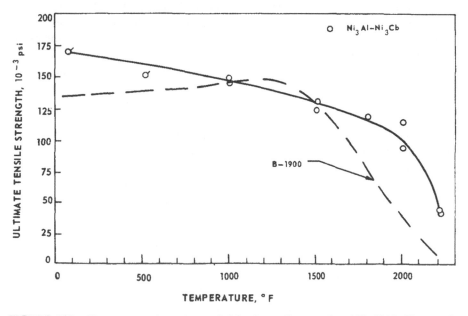

FIGURE 150. Temperature dependence of ultimate tensile strengths of Ni₃ Al-Ni₃ Cb eutectic and conventionally cast B-1900.

TABLE 21

Monotonic Properties of the Ni-Ni$_3$N$_b$ Composite

Properties	Average results	Range of results	No. of tests
Ultimate tensile strength, ksi	108.7	104.7 to 115.0	9
Strain at tensile fracture, %	12.4	11.25 to 15	4
Reduction in area, %	13.4	11.7 to 16.8	3
Notched tensile strength, ksi	139.8	138.0 to 143.4	3
Ultimate compressive strength, ksi	234.7	222.3 to 254.1	7
Compressive strain at ultimate, %	6.03	4.96 to 6.89	7

impact values than composites with needle-like whiskers. This contrasts to tensile behavior where the whisker morphology makes little difference. The Al-Al$_3$Ni composite exhibits fairly good toughness and low notch sensitivity despite the low tensile elongation of approximately 2% in the direction of the whisker reinforcement.[245]

The effects of cold rolling on the microstructure and properties of Ni-Al$_3$Ni eutectic composites have been reported.[246] The composites were readily rolled without damage to the filaments at reductions in thickness up to nearly 100% when rolled perpendicular to the Al$_3$Ni whisker alignment. When rolled parallel to the whisker direction, severe edge cracking occurred at very low reductions and the attempt was abandoned. The flat sheet specimens were initially 0.2 in. thick with the whiskers aligned in the plane of the sheet. Both the transverse and longitudinal strengths were improved by rolling with a maximum efficiency at approximately a 60% reduction in thickness. At this point no significant whisker breakage was noted.[246] The longitudinal and transverse tensile strengths were substantially increased to 50,000 and 18,000 psi, respectively. For a blade-like reinforcement morphology cold rolling was not as effective, increasing the longitudinal tensile strength to a maximum of 42,000 psi at 40% reduction in thickness. A fine dislocation substructure in the matrix produced during cold rolling disappears on annealing 1 hr at 500°C. The thermal stability of the microstructure at 610°C was substantially reduced compared to the unrolled composite. The preferred eutectic structure was drastically altered as whiskers agglomerated rapidly after heat treatment at 610°C. At 500°C no effect on microstructure was noted, but the properties of cold rolled material after annealing were essentially the same as the unrolled matrix. Intermediate anneals after partial cold rolling followed by further cold rolling also resulted in no improvement in properties over unrolled material. Thus, the cold rolling effect appears to be severely limited except as a means to produce thin sheet material of Al-Al$_3$Ni. It should be noted that the Al-Al$_3$Ni composites grown in this study had an initial tensile strength of 35,000 psi or less,[246] although utilizing the same techniques as previously employed,[247] where tensile strengths of 43,000 psi were reported. This is a substantial reduction in tensile strength four years later, which may indicate experimental problems in optimizing growth conditions to achieve reproducible microstructures.

The tensile strengths of several of the metal-monocarbide eutectic composites as a function of temperature in the uniaxial direction are indicated in Figure 151.[240] The data on the carbide eutectic composites are derived from several references,[240,248] as are the comparative data on TD nickel.[249] The tensile strengths are seen to be rather promising over the entire temperature range for these systems. During failure in tension a considerable number of whiskers were broken into short lengths even at higher temperatures. The measured elongations increased with increasing temperature and were higher than for TD nickel at elevated temperatures. Stress rupture tests were conducted on Ni-NbC specimens with intercarbide spacings of 7.5 μm, obtained by a 2.8 cm/hr growth rate. The results are shown in Table 22 and compare favorably with the stress rupture data for TD nickel and INCO®700 alloys.[240] The tensile ductility in the Ni-NbC system is considerably greater than that observed in many other eutectic composite systems such as Al-Al$_3$Ni.[240] In the

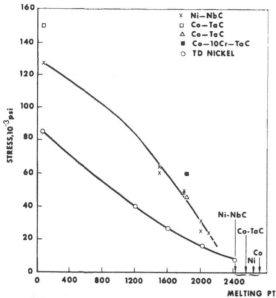

FIGURE 151. Temperature, °F tensile strengths of metal-monocarbide eutectics.

Co-TaC system[248] a substantial increase in strength has been achieved by adding 10 wt % Cr to the system without changing the volume fraction or composition of the TaC phase. Here again there is need for more characterization of off-axis properties such as the transverse strength and impact resistance.

The tensile and creep behavior of the (Co,Cr)-(Cr,Co)$_7$C$_3$ complex eutectic composite system have been studied.[250] The composite was found

TABLE 22

Ni-NbC Stress Rupture Data in the Longitudinal Direction

Temp. (°C)	Stress	Time to rupture (hr)	Total elong. %
815	30,000	98.4	–
981	25,000	3.8	17.5
981	15,000	132.8	2.9
1093	15,000	6.1	–
1093	10,000	19.8	–
1093	10,000	14.4	40.4
1093	10,000	9.5	29.7
1093	8,000	109.2	6.4
1093	5,000	>258.7*	–

*Ruptured in 8.0 hr after reloading to 10,000 psi.

to be anisotropic with considerable strengthening in the longitudinal direction and with transverse and 45° off-axis strengths limited by cracks initiating in the carbide phase at temperatures below 1800°F (981°C). From 1800° to 2200°F (1204°C) failure also occurred through matrix and interface cracking. The elastic modulus shows a marked temperature dependence but is much higher in the eutectic than in cobalt alloys because of the high modulus carbide phase. The longitudinally oriented creep behavior also exceeds that for the base alloy. The longitudinal strength approached 200,000 psi for the eutectic composite and dropped sharply at temperatures above 1300°F (704°C). At room temperature the 45° off-axis and transverse strengths are about 110,000 to 120,000 psi, which is close to the value for the unreinforced matrix. This must be considered reasonably good compared to the results in some of the other metal composite systems. Above 1500°F (815°C), the off-axis strengths also drop sharply. At temperatures of 2200°F (1204°C) the longitudinal, transverse, and unreinforced matrix strengths converge at well below 50,000 psi.

The effect of thermally induced stresses on the yield behavior of the (Co, Cr)-(Cr, Co)$_7$C$_3$ eutectic composite has been studied. An analysis to predict the observed yielding behavior has been elucidated based on the thermally induced residual stresses.[251] The difference in yield stress between tension and compression was predicted to decrease with increasing temperature and is found experimentally to do so with convergence at approximately 800°C. The stress relaxation temperature is found to be between 700° and 800°C compared to a calculated value of 950°C which is reasonable agreement considering that it is only a rough approximation.

The anisotropic toughness of the (Co,Cr)-(Cr,Co)$_7$C$_3$ eutectic system has also been evaluated.[252] The toughness was found to be anisotropic with the highest values obtained with the load deflection perpendicular to the fibrous reinforcement. The intermediate values were obtained where the three-point bend knife edges were parallel to the fibrous alignment. The lowest work to fracture was obtained from loads applied on the matrix with the fibers on end where cracks propagated easily along the fiber direction. Eutectics grown at 3 cm/hr were tougher than those grown at 2 cm/hr regardless of orientation. The spread in work of fracture varied from 1.6 to

7.2 ft-lb/in^2 or by a factor of four. The orientation giving the highest toughness values is the one normally utilized in experimental work on eutectic composites.

The compressive stress-strain behavior to failure of Al-CuAl$_2$ eutectic composites of small interlamellar spacings has been investigated.[253] Evidence of a dislocation-interface reaction was found in the deformation substructure of the aluminum-rich phase. Failure was found to be controlled by shear mode buckling of the lamellar structure. Buckling leads to CuAl$_2$ phase cleavage, aluminum phase shear, with void formation, coalescence, and crack growth. The 0.2% offset compressive yield strength ranged between 46 and 70,000 psi, and the strain to failure from 1.3 to 1.9%. The mechanical properties of a Ni-Cr eutectic composite were also determined in compression.[254] The 0.2% offset yield strength was determined at temperatures of −196°C to 850°C. The values dropped from 160,000 psi to 35,000 psi over this range. Swaging to a reduction in area of 40% slightly increased the values at intermediate temperatures and, when followed by a one half hr anneal at 1000°C, raised the values to 260,000 psi at −196°C, dropping again to 35,000 psi at 850°C. A compressive yield strength of 275,000 psi has been attained in a Ti-Ti$_5$Si$_3$ eutectic composite after heat treatment.[255]

A different type of property measurement should become of increasing interest as the directional electrical, magnetic, and optical properties of eutectic composite structures are more fully studied. A review of electromagnetic properties of eutectic composites has recently been completed by Weiss.[203] The system InSb-NiSb, which consists of rods or needles of NiSb in an InSb matrix, is considered by Weiss to be the outstanding system for marketable devices. A number of other potential systems have also been mentioned by Galasso.[24] Eutectic composites have been produced in the Co-Nb and Co-Nb-Fe systems which consist of lamellar NbCo$_3$ in a Co-rich matrix. Marked changes in magnetic properties of eutectic composites were observed with changes in composition.[256]

Chapter 14

MECHANICS OF COMPOSITES

Introduction

This section will serve to introduce the reader to some of the problems and approaches that can be used to understand and predict the mechanical behavior of composites. The study of mechanical behavior of metal matrix composites largely has been confined to simple property measurements and test configurations. As a consequence this is more a review of approaches that can be employed than an account of results that have been obtained.

Sendeckyj[257] views the ability to predict the strength of composite materials under complex loading conditions in terms of two basic approaches: empirical and micromechanics. In an empirical theory the strength criterion is assumed consistent with broad concepts of materials behavior. Characteristic parameters of the criterion are evaluated by "simple" tests and the applicability of the criterion is verified by running other tests. At best, these tests show that the theory is applicable under specific loading conditions. Experimentally, one can show only that a criterion is invalid when the predicted strengths are not observed. A large amount of testing will not prove the validity of the theory although it will lend credibility to the strength criterion. In micromechanics, the failure criterion is derived from first principles by postulating failure modes and employing "rigorous" analyses to predict the strength. The results are then compared to experimental data. Sendeckyj does not feel this approach has yet led to a satisfactory strength criterion.

Pagano and Tsai[258] make the distinction between micromechanics and macromechanics as related to the analytical investigation of composites where micromechanics considers the true identity of a composite by recognizing the constituent fiber and matrix phases present. Solutions are derived for a heterogeneous material with assumed continuity conditions involving the bonding of the matrix to the fiber. Coupled with mathematical analyses, the constitutive equations for the composite as a whole can be predicted. This makes possible the subsequent treatment of unidirectional composites as macroscopically homogeneous bodies.

Macromechanics defines the response of a structure in terms of individual layers or plies. For layered or laminated structures, classical laminated plate or shell theory is used. The relation between micro and macromechanics is thus established as the dependence of the response of structural members to the mechanical properties of the constituents of the composites. Laminate theory is popular because organic matrix composite structures are often fabricated by a lamination process in which plies of oriented parallel filaments are stacked in some predetermined stacking sequence.

Before proceeding further, a few definitions should be considered. First, a filamentary composite is one in which continuous filaments rather than short whiskers or chopped fibers are used. This generally implies that the filament traverses from one boundary to another in the composite in some oriented manner. An array of one row in-plane of parallel filaments surrounded by matrix material is a single ply or lamina. The axes which are parallel and perpendicular to the filament direction are the lamina principal axes. In conventional terms a lamina as described would be anisotropic and heterogeneous. In composite mechanics, since the properties along a given axis are uniform from point to point, the lamina is homogeneous. It is orthotropic because there are three mutually perpendicular planes of material property symmetry which can be passed through a point. In an isotropic material, all planes passing through a point are planes of material, property symmetry. For anisotropic materials, there are no planes of material property symmetry which pass through a point. A constitutive equation is one in which the Hooke's law (stress-strain) relationships are used to describe the mechanical properties of a material. For a more detailed review of these terms, the monograph of Ashton, Halpin, and Petit[259] should be consulted.

The importance of considering the anisotropic response of composite materials to complex loading conditions has been emphasized by Halpin, Pagano, Whitney, and Wu.[260] They demonstrated the necessity to revise classical methods of mechanical characterization of isotropic solids to determine the stiffness and fracture properties of anisotropic solids. An experimental procedure was employed involving testing with thin walled anisotropic cylinders for mechanical characterization. More recently, the use of thin walled cylinders for composite testing has been reviewed and the importance of maintaining a large radius to thickness ratio in the test specimens has been emphasized.[261] A serious problem is the buckling failure that arises as the ratio increases. With metal matrix composites there is also the practical experimental difficulty of fabricating thin walled tubes with uniform wall thickness and an invariant number of plies.

Before considering the more complex methods of prediction of moduli and strengths and the mathematical characterization of the response of a composite material to various loading environments, the often used rule of mixtures needs consideration. It is the approach generally employed to determine whether a metal matrix composite has interesting properties initially, and as a measure of optimizing fabrication parameters or bonding conditions. Much of the following review is based on a discussion by B. R. Collins in a previous review of mechanical properties.[262] Somewhat more detail will be found in that reference.

Rule of Mixtures

Predictions of the response of a unidirectionally reinforced composite were based initially upon postulated states of stress and load transfer mechanisms. The outgrowth of these analyses was the rule of mixtures. By assuming an isostrain criterion, i.e., both fiber and matrix are strained equally and uniformly, the longitudinal and transverse strength, stiffness, and the major and minor Poisson's ratio were found by the use of parallel and series spring models. The assumptions are a. elastic (and plastic) isotropy; b. perfect mechanics continuum at interfaces; c. no chemical reaction between constituents; d. an absence of residual stresses; e. an absence of rheological interaction at interfaces; f. definability of in situ properties of both fiber and matrix properties at the strains in question. Experimental data have shown that only the longitudinal modulus and major Poisson's ratio can be reliably predicted by this simple approximation. Kelly and Davies[174] reported these approximations as

$$E_{11} = E_F V_F + E_m(1-V_F) \tag{1}$$

and

$$\nu_{12} = \nu_f V_f + \nu_m(1-V_f) \tag{2}$$

where

E_{11} = Young's modulus of the composite parallel to the fibers
E_f = Young's modulus of the fiber
E_m = Young's modulus of the matrix
ν_{12} = Major Poisson's ratio of the composite
ν_f = Poisson's ratio of the fiber
ν_m = Poisson's ratio of the matrix
V_f = Vol % of fiber in the composite

Experimental data[174] fit these moduli approximations well, but Hill[263] showed, theoretically, that these predictions were really the lower bound to the moduli and applicable when the Poisson's ratios of the constituents were equal. The moduli prediction can be extended to the region where the matrix has yielded by substituting for E_m and ν_m in Equations 1 and 2:

$$E_{11} = E_f V_f + \left(\frac{d\sigma_m}{d\epsilon_m}\right)\bigg|_\epsilon (1-V_f) \tag{3}$$

$$\nu_{12} = \nu_f V_f + \nu_m\big|_\epsilon (1-V_f) \tag{4}$$

where $\frac{d\sigma_m}{d\epsilon_m}\big|_\epsilon$ is the slope of the stress-strain curve of the matrix material free of fibers, at the equivalent strain of the composite, and $\nu_m\big|_\epsilon$ is the Poisson's ratio of the matrix at the same strain varying from the elastic value to 0.5 for an ideally plastic material.

According to the rule of mixtures prediction, the longitudinal strength of the composite would be

$$\sigma_{11} = \sigma_f V_f + \sigma_m(1-V_f) \tag{5}$$

where

σ_f = stress in the fiber

and

σ_m = stress, at the fracture strain of the composite, in the matrix.

This prediction of longitudinal strength in Equation 5 is the one that has proven so useful in practice. The previous equations give properties such as the modulus which are rather insensitive to important factors such as the nature of the filament-matrix interface. They are thus of much less utility in studying composite behavior. It should be emphasized that while this simple *strength prediction correlates well with experimental data, there is no reason to expect it to be highly accurate. Fabrication problems would be expected to decrease composite strength. This would include factors such as misalignment of fibers, cracking of some brittle fibers during bonding, and degradation through chemical reaction. Zweben[264] points out that the fibers generally used in composites do not have uniform ultimate strengths so there is no reason to expect the rule of mixtures to apply. He also notes that since the fibers are brittle they are sensitive to surface flaws and other defects and, therefore, the mean filament strengths should decrease with increasing gage length. Rosen[265] used the idea of the bundle strength of the filaments to determine the strength which could be used for σ_f in Equation 5.

Prediction of Elastic Constants

It has been shown that E_1 and ν_{12} can be accurately predicted by the rule of mixtures. However, for an orthotropic material subjected to plane stress four moduli are required to relate stress and strain (five constants exist, but only four are independent). More exact analyses are necessary for the reliable prediction of these three remaining constants. Rigorous solutions for E_1 and ν_{12} are also included in these formal approaches.

Ashton, Halpin, and Petit[259] divide these approaches into three categories: 1. self-consistent model methods, 2. variational methods, and 3. exact methods, which have the following assumptions common to all:

1. The ply is macroscopically homogeneous, linearly elastic, and generally orthotropic or transversely isotropic.

2. The fibers are linearly elastic and homogeneous.

3. The matrix is linearly elastic and homogeneous.

4. Fiber and matrix are free of voids.

5. There is complete bonding at the interface of the constituents and there is no transition region between them.

6. The ply is initially in a stress free state.
7. The fibers are
   a. regularly spaced
   b. aligned

It is apparent from the assumptions that these analyses can be rigorous only in the mathematical sense, since the assumptions are in no instance realized. For most metallic composites these assumptions are wrong. For example, residual stresses are common within the plies, and if there is no transition between the fibers and the matrix, there probably is no bond metallurgically at all.

The following review of these methods has been adapted in part from Ashton, Halpin, and Petit,[259] Pagano and Tsai[258] and Lin.[266]

In the Self-Consistent Model Method a fiber and matrix simulator is surrounded by an unbounded medium which is macroscopically indistinguishable from the composite.[267] If the medium is subjected to a uniform load applied at infinity, a uniform strain field is induced in the filament[268] and the model can be analyzed. Using this method, Hill[269] derived estimates for all the elastic constants and subsequently showed that the expressions were reliable for low filament volume ratios, reasonable at intermediate values, and unreliable at high ratios. Kilchinskii[270] and Hermans[267] varied the approach of Hill by surrounding the fiber by a concentric cylinder and ascribed to the cylinder the properties of the matrix. Kilchinskii assumed a hexagonal array distribution of filaments in the matrix and modeled the system by requiring the outer radius of the matrix cylinder to be such that the cross-sectional areas of the cylinder and hexagonal unit were the same. Hermans merely required that the ratio of the volume of the fiber to the matrix cylinder be equal to the volume of fiber in the composite without stipulating the regular spacing of the fibers. Whitney and Riley[271] modeled a similar system by imbedding a filament in a cylinder of finite radius to obtain reasonable results for the normal and shear moduli.

The Variation Method employs energy theorems of classical elasticity to obtain limiting bounds on the elastic constants of the orthotropic body or lamina. The lower bound is obtained from the minimum complementary energy theorem and the upper bound is obtained from the minimum potential energy theorem.[272] Paul[273] and Hill[263] used this method to show that the rule of mixtures prediction of E_2:

$$\frac{1}{E_2} = \frac{V_F}{E_F} + \frac{1-V_F}{E_m} \qquad (6)$$

was the lower bound for a macroscopically isotropic composite. Bounds for G_{23} and ν_{23} were also presented by Paul. Hashin and Rosen[274] assumed a lamina to be transversely isotropic and derived upper and lower bounds for all five elastic constants for hexagonal and random arrays of fibers. This analysis was extended by Hashin[275] to obtain bounds for the bulk and two shear moduli of material with constituents of arbitrary phase geometry. In most cases these bounds are far apart and do not compare favorably with experimental data.

If the fibers are arranged in the composite in a periodic, geometric array, elastic fields can be developed by series development and numerical techniques. Expressions for the elastic constants can be obtained by averaging these elastic fields. This approach is the Exact Method. Inherent in the solutions are the assumptions that the fibers are isotropic and form a doubly periodic array. By using symmetry arguments to reduce the problem to a finite region, called the fundamental repeating element, an alternative approach exists for the solution of the elastic constants by this method. Van Fo Fy[276] derived approximate and exact solutions for all the elastic constants from a series development. Herman and Pister[277] reduced the problem to a square with a rigid circular inclusion and presented results for the elastic constants, two coefficients of linear thermal expansion, and two heat conductivities, with a numerical scheme. Wilson and Hill[278] used a rectangular fundamental element with a hole in the center and solved the elasticity problem by complex variables and by mapping the element into two concentric cylinders. Adams and Bloom[279] used a hexagonal array to solve for E_1 and ν_{12} which substantiated the accuracy of Equations 1 and 2. Adams et al.[280-282] used a finite difference technique on a rectangular element which yielded values of E_2 above the upper bound of Hashin and Rosen.[274] Experimental results on organic composites showed this prediction to be quite accurate. G_{12} was shown to be very close to the upper bound. Pickett[283,284] used a series solution in polar coordinates and satisfied the boundary conditions at the fiber-matrix interface and the sides of the fundamental element. He used a hexagonal array

reduced to a triangular element and obtained solutions within the bounds of Hashin and Rosen. Bloom and Wilson[285] also reduced a hexagonal array to a triangular element and, by selecting an infinite series which satisfied the boundary conditions at the fiber-matrix interface exactly, obtained good results for E_1 and G_{12}. Foye[286,287] applied a discrete element method to predict E_2, G_{12}, ν_{12} and ν_{23} for square, diamond, and hexagonal arrays. Solutions for the hexagonal array lie within the bounds of Hashin and Rosen for E_2 while the square array yields solutions above those bounds which fit better with experimental data. G_{12} in Foye's analyses fits the hexagonal array very well.

Because the approaches have been rigorous and have considered fiber-fiber interaction, they are expected to have some validity for metal composites. These solutions, according to Tsai,[288] should preempt the idealizations of netting analysis, shear lag analysis, and strength of materials. Tsai points out that the netting analysis ignores the matrix and governing equations of elasticity, the shear lag analysis ignores fiber-fiber interaction, and strength of materials misrepresents the composite by uniaxial tensile bars connected in series or parallel or a combination of the two.

To overcome the shortcomings of these solutions, some investigators have attempted to completely characterize unidirectional composites on a microstress level, recognizing residual stresses and fiber-matrix interaction.

Application of these solutions is cumbersome and limited. Ashton, Halpin, and Petit[259] note that limitations arise because necessary curves are available for only a few material parameters. To facilitate simple estimates of lamina properties, Halpin and Tsai[289] have developed approximate formulas capable of interpolating the exact machine calculations available. They used Hermans' solution[267] generalizing Hill's self-consistent model[269] to reduce the equations to the following form:

$$E_1 \cong E_F V_f + E_m V_m$$

$$\nu_{12} \cong \nu_f V_f + \nu_m V_m$$

and

$$\frac{\bar{P}}{P_m} = \frac{(1 + \zeta \eta V_f)}{(1 - \eta V_f)} \qquad (7)$$

where

$$\eta = (P_f/P_m - 1) / (P_f/P_m + \zeta) \qquad (8)$$

and

$\overline{P}$ = composite moduli, E_2, G_{12}, υ_{23}

P_f = corresponding fiber modulus, E_f, G_f, υ_f, respectively

P_m = corresponding matrix modulus, E_m, G_m, υ_m

ζ = a measure of reinforcement which depends on the boundary conditions

Once the ζ factors are known for the geometry of inclusions, packing geometry, and loading conditions, the composite elastic moduli for fibrous composites are approximated from the Halpin-Tsai equations. Reliable estimates for the ζ factor are obtained by comparing Equations 7 and 8 with the micromechanics solutions of formal elasticity theory. The validity of these equations is shown by Ashton, Halpin, and Petit[259] by comparing the predictions of the Halpin-Tsai equations with results obtained by Adams and Doner,[280,281] as done in Figure 152.

For square fibers the ζ factor was found to be 1 for G_{12} and 2 for E_2.[286,287] It was also found that the factors were functions of width to thickness ratios of the form

$$\zeta_{E_2} = 2(a/b)$$
$$\log \zeta_{G_{12}} = \sqrt{3} \log (a/b) \qquad (9)$$

Approximate predictions of all elastic constants are now readily available for unidirectional composites.

Laminate Constitutive Equations

Once the elastic constants of the orthotropic lamina are known (either predicted or measured) the constitutive relationships of the ply can be developed. This is illustrated by an abbreviated development of these relationships for an anisotropic material to the specialized case of plane stress of an orthotropic lamina.

For a homogeneous, isotropic material in a one-dimensional stress-state, Hooke's law relationship is

$$\sigma = E\epsilon \qquad (10)$$

where the proportionality constant E is the modulus of elasticity. Only one elastic constant is necessary to describe the stress-strain relationship. For the same isotropic material in a two-dimensional stress-state, the number of independent elastic constants becomes 2, i.e.,

$$\sigma_1 = (\epsilon_1 + \mu\epsilon_2) \frac{E}{1 - \mu^2}$$
$$\sigma_2 = (\epsilon_2 + \mu\epsilon_1) \frac{E}{1 - \mu^2} \qquad (11)$$
$$\tau_{12} = \gamma_{12} G$$

where

$$G = \frac{E}{2(1 + \mu)}$$

and the elastic constants are

E = modulus of elasticity
G = shear modulus
μ = Poisson's ratio

The generalized case for an anisotropic material is written in tensor notation as

$$\sigma_{ij} = C_{ijkl} \epsilon_{kl}$$
or
$$\epsilon_{ij} = S_{ijkl} \sigma_{kl} \qquad (12)$$

where:

σ_{ij} = stress components
ϵ_{ij} = strain components
C_{ijkl} = stiffness coefficients
S_{ijkl} = compliance coefficients

In expanded notation from Equation 12 one obtains 81 elastic constants for the general case. Tsai[290] has shown that $[\epsilon]$, $[\sigma]$, and $[S]$ must be symmetric and that Equation 13 can be written in contracted form for an orthotropic material as

$$
\begin{Bmatrix} \epsilon_1 \\ \epsilon_2 \\ \epsilon_3 \\ \gamma_{23} \\ \gamma_{13} \\ \gamma_{12} \end{Bmatrix}
=
\begin{bmatrix}
S_{11} & S_{12} & S_{13} & 0 & 0 & 0 \\
S_{12} & S_{22} & S_{23} & 0 & 0 & 0 \\
S_{13} & S_{23} & S_{33} & 0 & 0 & 0 \\
0 & 0 & 0 & S_{44} & 0 & 0 \\
0 & 0 & 0 & 0 & S_{55} & 0 \\
0 & 0 & 0 & 0 & 0 & S_{66}
\end{bmatrix}
\begin{Bmatrix} \sigma_1 \\ \sigma_2 \\ \sigma_3 \\ \tau_{23} \\ \tau_{13} \\ \tau_{12} \end{Bmatrix}
\qquad (13)
$$

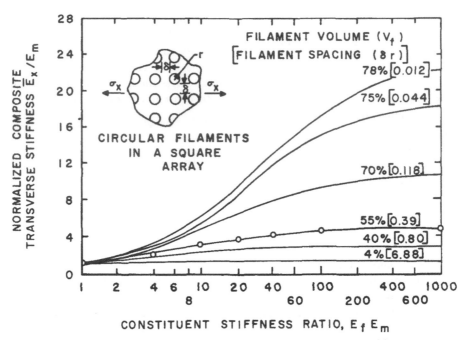

FIGURE 152A. Comparison of Halpin-Tsai calculation, solid circles, with square array calculations for transverse stiffness.

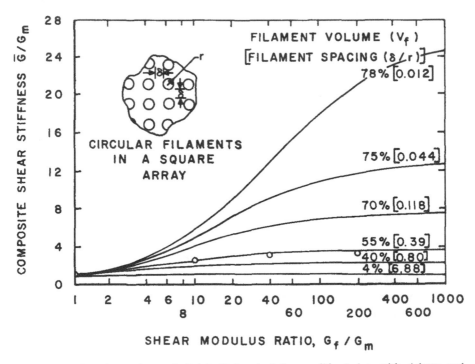

FIGURE 152B. Comparison of Halpin-Tsai calculation, solid circles, with Adams and Donner's square array calculations for longitudinal shear stiffness.

where

$$\varepsilon_{11} = \varepsilon_1 \qquad\qquad \sigma_{11} = \sigma_1$$

$$\varepsilon_{22} = \varepsilon_2 \qquad\qquad \sigma_{22} = \sigma_2$$

$$\varepsilon_{33} = \varepsilon_3 \qquad\qquad \sigma_{33} = \sigma_3$$

$$\varepsilon_{23} = \gamma_{23} \qquad\qquad \sigma_{23} = \tau_{23}$$

$$\varepsilon_{13} = \gamma_{13} \qquad\qquad \sigma_{13} = \tau_{13}$$

$$\varepsilon_{12} = \gamma_{12} \qquad\qquad \sigma_{12} = \tau_{12}$$

Equation 13 is further reduced for the case of plane stress

$$\sigma_3 = \tau_{23} = \tau_{13} = 0$$

to the form

$$\begin{Bmatrix} \varepsilon_1 \\ \varepsilon_2 \\ \gamma_{12} \end{Bmatrix} = \begin{bmatrix} S_{11} & S_{12} & 0 \\ S_{12} & S_{22} & 0 \\ 0 & 0 & S_{66} \end{bmatrix} \begin{Bmatrix} \sigma_1 \\ \sigma_2 \\ \tau_{12} \end{Bmatrix} \qquad (14)$$

Where now only four independent elastic constants are necessary to specify the response of the lamina, and the components of the [S] matrix can be written in terms of the standard engineering constants as

$$S_{11} = \frac{1}{E_1}$$

$$S_{22} = \frac{1}{E_2}$$

$$S_{12} = \frac{\mu_{12}}{E_1} = -\frac{\mu_{21}}{E_2} \qquad (15)$$

$$S_{66} = \frac{1}{G_{12}}$$

$$S_{16} = S_{26} = 0$$

It is common to write Equation 14 in terms of the reduced stiffness matrix [Q] where [Q] = [S]$^{-1}$ so that

$$\begin{Bmatrix} \sigma_1 \\ \sigma_2 \\ \tau_{12} \end{Bmatrix} = \begin{bmatrix} Q_{11} & Q_{12} & 0 \\ Q_{12} & Q_{22} & 0 \\ 0 & 0 & 2Q_{66} \end{bmatrix} \begin{Bmatrix} \varepsilon_1 \\ \varepsilon_2 \\ \frac{1}{2}\gamma_{12} \end{Bmatrix} \qquad (16)$$

The preceding discussion assumed the natural axes of the lamina were coincident with the reference axes. If the two axes systems vary by some angle θ, the constitutive relationships can be transformed to the reference axes by using the usual standard transformation matrix for plane stress [T] where

$$[T] = \begin{pmatrix} m^2 & n^2 & mn \\ n^2 & m^2 & -2mn \\ -mn & mn & m^2-n^2 \end{pmatrix} \qquad (17)$$

and $m = \cos\theta$

$n = \sin\theta$

If [Q] is written with respect to the reference axes Equation 16 becomes

$$\begin{Bmatrix} \sigma_x \\ \sigma_y \\ \tau_{xy} \end{Bmatrix} = [\bar{Q}] \begin{Bmatrix} \varepsilon_x \\ \varepsilon_y \\ \gamma_{xy} \end{Bmatrix} \qquad (18)$$

where

$$\bar{Q}_{11} = Q_{11}\cos^4\theta + 2(Q_{12} + 2Q_{66})\sin^2\theta\cos^2\theta$$
$$+ Q_{22}\sin^4\theta$$

$$\bar{Q}_{22} = Q_{11}\sin^4\theta + 2(Q_{12} + 2Q_{66})\sin^2\theta\cos^2\theta$$
$$+ Q_{22}\cos^4\theta$$

$$\bar{Q}_{12} = (Q_{11} + Q_{22} - 4Q_{66})\sin^2\theta\cos^2\theta$$
$$+ Q_{12}(\sin^4\theta + \cos^4\theta)$$

$$\bar{Q}_{66} = (Q_{11} + Q_{22} - 2Q_{12} - 2Q_{66})\sin^2\theta\cos^2\theta$$
$$+ Q_{66}(\sin^4\theta + \cos^4\theta)$$

$$\bar{Q}_{16} = (Q_{11} - Q_{12} - 2Q_{66})\sin\theta\cos^3\theta$$
$$+ (Q_{12} - Q_{22} + 2Q_{66})\sin^3\theta\cos\theta$$

$$\bar{Q}_{26} = (Q_{11} - Q_{12} - 2Q_{66})\sin^3\theta\cos\theta$$
$$+ (Q_{12} - Q_{22} + 2Q_{66})\sin\theta\cos^3\theta$$

$$(19)$$

The stress-strain relations are now determined for a generally orthotropic material referred to any coordinate axis. Equation 19 also describes the variation of the stiffness matrix as it is rotated through an angle θ. The entire elastic behavior of undirectional composites subjected to plane stress is thus defined.

If individual lamina are now stacked to form a plate or shell, the response can be characterized by considering the contribution of each lamina. For this development the strain matrix $[\epsilon]$ is divided into two components:

$$
\begin{Bmatrix} \epsilon_x \\ \epsilon_y \\ \gamma_{xy} \end{Bmatrix} = \begin{Bmatrix} \epsilon_x^o \\ \epsilon_y^o \\ \gamma_{xy}^o \end{Bmatrix} + Z \begin{Bmatrix} k_x \\ k_y \\ k_{xy} \end{Bmatrix}
\tag{20}
$$

where ϵ_x, ϵ_y, and γ_{xy} are midplane strains, k_x, k_y, and k_{xy} are the lamina curvatures, and Z is the distance from the midplane to the point where the stress is being determined. By defining a stress and moment resultant as a system of stresses and moments statically equivalent to the stress system applied to the laminate but applied at the laminate midplane, Equation 19 can be written in two parts as

$$
\begin{Bmatrix} N_x \\ N_y \\ N_{xy} \end{Bmatrix} = \begin{bmatrix} A_{11} & A_{12} & A_{16} \\ A_{12} & A_{22} & A_{26} \\ A_{16} & A_{26} & A_{66} \end{bmatrix} \begin{Bmatrix} \epsilon_x^o \\ \epsilon_y^o \\ \gamma_{xy}^o \end{Bmatrix} + \begin{bmatrix} B_{11} & B_{12} & B_{16} \\ B_{12} & B_{22} & B_{26} \\ B_{16} & B_{26} & B_{66} \end{bmatrix} \begin{Bmatrix} k_x \\ k_y \\ k_{xy} \end{Bmatrix}
\tag{21}
$$

$$
\begin{Bmatrix} M_x \\ M_y \\ M_{xy} \end{Bmatrix} = \begin{bmatrix} B_{11} & B_{12} & B_{16} \\ B_{12} & B_{22} & B_{26} \\ B_{16} & B_{26} & B_{66} \end{bmatrix} \begin{Bmatrix} \epsilon_x \\ \epsilon_y \\ \gamma_{xy}^o \end{Bmatrix} + \begin{bmatrix} D_{11} & D_{12} & D_{16} \\ D_{12} & D_{22} & D_{26} \\ D_{16} & D_{26} & D_{66} \end{bmatrix} \begin{Bmatrix} k_x \\ k_y \\ k_{xy} \end{Bmatrix}
\tag{22}
$$

where

$$
N_x = \int_{-h/2}^{h/2} \sigma_x dz \qquad M_x = \int_{-h/2}^{h/2} \sigma_x dz
$$

$$
N_y = \int_{-h/2}^{h/2} \sigma_y dz \qquad M_y = \int_{-h/2}^{h/2} \sigma_y z dz
\tag{23}
$$

$$
N_{xy} = \int_{-h/2}^{h/2} \tau_{xy} dz \qquad M_{xy} = \int_{-h/2}^{h/2} \tau_{xy} z dz
$$

and

$$
A_{ij} = \sum_{k=1}^{n} (\bar{Q}_{ij})_k (h_k - h_{k-1})
$$

$$
B_{ij} = \sum_{k=1}^{n} (\bar{Q}_{ij})_k (h_k^2 - h_{k-1}^2)
\tag{24}
$$

$$
D_{ij} = \frac{1}{3} \sum_{k=1}^{n} (\bar{Q}_{ij})_k (h_k^3 - h_{k-1}^3)
$$

where k is a dummy variable representing each lamina and n is the number of plies.

Equations 21 and 22 can be combined to form the total constitutive equation for an anisotropic plate [259]

$$
\begin{Bmatrix} N \\ \hline M \end{Bmatrix} = \begin{bmatrix} A & | & B \\ \hline B & | & D \end{bmatrix} \begin{Bmatrix} \epsilon^o \\ \hline k \end{Bmatrix}
\tag{25}
$$

Once the constitutive relationships of the laminate are determined, the governing equations of plates and shells can be derived. Only a brief development follows; for specific solutions several references are available.[259,266,291-293] These equations have the same limitations as bending of ordinary plates and apply only for small deflections.

The equations of equilibrium for laminated plates are identical to those of homogeneous materials.[259] For a thin plate these are

$$
\frac{\partial N_x}{\partial x} + \frac{\partial N_{xy}}{\partial y} = 0
\tag{26}
$$

$$
\frac{\partial N_y}{\partial y} + \frac{\partial N_{xy}}{\partial x} = 0
\tag{27}
$$

$$
\frac{\partial^2 M_x}{\partial x^2} + \frac{2 \partial^2 M_{xy}}{\partial x \partial y} + \frac{\partial^2 M_y}{\partial x^2} = -q(x,y)
\tag{28}
$$

where N_x, N_y, N_{xy} are the stress resultants, M_x, M_y, M_{xy} are the moment resultants, and $q(x,y)$ is the transverse loading distributed on the plate. Equations 26 and 27 can be identically satisfied by employing the Airy Stress function such that

$$
N_x = \frac{\partial^2 U}{\partial y^2}
$$

$$
N_y = \frac{\partial^2 U}{\partial x^2}
\tag{29}
$$

$$
N_{xy} = \frac{\partial^2 U}{\partial x \partial y}
$$

By rewriting the equation $[M] = [B][\epsilon^o] + [D][k]$ in the form

$$
[M] = [C^*][N] + [D^*][k]
\tag{30}
$$

where

$$
[C^*] = [B][A]^{-1}
$$

$$
[D^*] = [D] - [B][A]^{-1}[B]
$$

One governing equation for laminated plates is developed in complex form. The second necessary complex equation comes from the compatibility condition of midplane strain.

These two equations (not given — see Reference 262) make it possible, in theory, to solve for the stress function U and the deflection w, subject to appropriate boundary conditions. As in the case of the development of the laminate constitutive equations, advantage is taken of symmetry to simplify the equations. For example, if a plate has the orientation of plies symmetric about the midplane of the laminate, the coupling matrix [B] becomes equal to zero. As a direct result [B*] = [C*] = [0] and [D*] = [D].

If simplications due to midplane symmetry are also considered, and if the laminate is orthotropic so that $A_{16} = A_{26} = 0$ the complex equations reduce to

$$A_{11}^* \frac{\partial^4 U}{\partial y^4} + (2A_{12}^* + A_{66}^*) \frac{\partial^4 U}{\partial x^2 \partial y^2} + A_{22}^* \frac{\partial^4 U}{\partial x^2} = 0 \tag{31}$$

and

$$D_{11} \frac{\partial^4 w}{\partial x^4} + (2D_{12} + D_{66}) \frac{\partial^4 w}{\partial x^2 \partial y^2} + D_{22} \frac{\partial^4 w}{\partial y^4} = q(x,y) \tag{32}$$

These equations can be solved by the same methods used to obtain solutions for isotropic plates.

Complex Stress States and Strength Predictions

In metal-matrix systems, unidirectionally reinforced composites can sometimes be used since off-axis loads can be sustained by the metal matrix. The emphasis of stiffness and strength predictions has, therefore, focused on these structures. The entire stress-strain curve is predicted for a longitudinal or transverse load if the mechanical properties of the constituents are known. The approach offers the advantage of considering the constituent materials of the composite individually. Besides considering unidirectional material, the subsequent discussion can also apply to a single lamina in a plate. In considering micromechanics strength predictions, it should be emphasized that none of these solutions has widespread acceptance in the field of composites. The discussion here differs from that on the prediction of elastic constants in that the complex stress-states induced in the composite upon loading are considered, while theoretical characterization is extended beyond yielding.

As noted by Taylor et al.,[294] Piehler[134] was one of the first to study the elastic interior stress field of a unidirectional composite under axial loading. He used a circular filament surrounded by a hexagonal matrix as the fundamental repeating element and analyzed the effects on the boundary by adjacent elements. He showed that the in-plane

stresses generated by axial loads were dependent upon the mismatch of Poisson's ratio of the constituent materials. For an epoxy-glass system which had similar Poisson's ratios the in-plane stresses were very small, while these stresses approached 3% of the axial stress for a silver-steel composite with close packed fibers.

Bloom and Wilson[285] used the same hexagonal element and, by using a plain-strain analysis, showed that for E_f/E_m ratios of 120, 20, and 6 (which represent the boron-resin, glass-epoxy, and boron-aluminum systems, respectively) and Poisson's ratios of 0.20 (boron) and 0.35 (aluminum), in-plane stresses were very low, even for high filament volumes.

Ebert and Gadd[77] developed a concentric binary cylinder model to investigate the response of a unidirectional composite to axial loading. By considering the von Mises and Tresca yield criteria, Ebert et al.[78] extended this model to the region where the matrix and/or fiber yielded. Also, by using the Prandtl-Reuss equations they were able to consider strain hardening of the matrix. In later papers[176,295] Ebert showed that residual stresses were quite severe, but the effects could be ameliorated by mechanically prestraining the composite and altering the triaxial stress state.

Haener and Ashbaugh[296] extended these analyses to include residual stresses induced in the composite when it cools down from the initial consolidation temperature. These stresses arise due to the mismatch of coefficients of thermal expansion.

Taylor, Dolowy, and Shimizu[297] extended the concept of constraints imposed by multiaxial stresses on a material to show that boron/aluminum composites could be expected to show moduli and strengths above those predicted by the rule of mixtures at strains lower than those necessary to develop the full strength of the boron filament. Fabrication stresses were considered in this development.

Foye[298] used the concept of stress concentrations introduced by rectangular inclusion arrays in the matrix to determine the state of stress and the strength of a composite.

Solutions which predict the strength and state of stress in a composite when loaded transverse to the direction of the fibers are not as numerous as the axial case. Chen and Lin[102] and Adams and Doner[281] theoretically predicted the effect of the volume fraction of fiber on the transverse modulus. By using the von Mises yield criterion, Chen and Lin were also able to predict the transverse strength as a function of filament content, provided failure occurred in the matrix. Kreider, Dardi, and Prewo[123] were able to show that experimental data on Borsic/Al fit the curves of Adams and Chen very well, despite matrix porosity, residual stresses, and filament splitting.

The semi-empirical alternative to micromechanics, which has been developed independently by Norris[299,300] and Hill,[301] determines the strengths along the principal axes and then obtains biaxial, or off-axis, strengths by some interpolation scheme. The strength theories are merely adaptations of theories for homogeneous, isotropic materials.[302] The most common criteria of a failure are 1. maximum normal stress, 2. distortional energy, 3. maximum shear stress, and 4. maximum strain.

The Lame-Navier maximum stress theory was modified by Jenkin[303] to predict the failure of an orthotropic material. The stresses were resolved into components along the natural axes of the material. Failure occurred when these resolved stresses exceeded the tensile or compressive strength of the material:

$$\sigma_1 = F_1$$
$$\sigma_2 = F_2 \tag{33}$$

Hill[301] generalized the distortional energy criterion (the von Mises yield criterion) of isotropic materials. Extending the criterion for isotropy to anisotropic bodies of assumed material symmetry, Hill developed a quadratic function for the "plastic potential,"

$$2f(\sigma_{ij}) = F(\sigma_{yy} - \sigma_{zz})^2 + G(\sigma_{zz} - \sigma_{xx})^2$$
$$+ H(\sigma_{xx} - \sigma_{yy})^2$$
$$+ 2L\,\tau_{yz}^2 + 2M\,\tau_{zx}^2 + 2N\,\tau_{xy}^2 = 1 \tag{34}$$

where F, G, H, L, M, and N are material coefficients characteristic of the state of anisotropy and x, y, and z are the axes of material symmetry which is assumed to exist, and

$$2F = (Y)^{-2} + (Z)^{-2} - (X)^{-2}$$
$$2G = (Z)^{-2} + (X)^{-2} - (Y)^{-2}$$
$$2H = (X)^{-2} + (Y)^{-2} - (Z)^{-2} \tag{35}$$
$$2L = (S_{23})^{-2}$$

$$2M = (S_{31})^{-2}$$

and

$$2N = (S_{12})^{-2}$$

X, Y, and Z are determined from either uniaxial tension or compression tests and S_{23}, S_{31}, and S_{12} are determined from pure shear tests in the yz, zx, and xy planes, respectively.

Marin[304] attempted to further generalize the von Mises theory to account for differences in a material's tensile and compressive yield strengths. He limited the theory to principal stresses. This restriction complicated the solution considerably because in an anisotropic body principal stress axes are not always coincident with principal strain axes.[305] Norris[299,300] used the natural axes of the material and proposed an "interaction" formula relating stress components which would cause failure to applied loads. In considering an orthotropic material resembling a waffle grid he developed three von Mises types of equations which reduced to the von Mises isotropic yield criteria if isotropic strength relations held. Azzi and Tsai[306] first applied Hill's criterion to unidirectional, fiber reinforced composites. By assuming the composite to be transversely isotropic and, therefore, setting G equal to H, they reduced the failure criterion to

$$\frac{\sigma_{xx}^2 - \sigma_{xx}\sigma_{yy}}{X^2} + \frac{\sigma_{yy}^2}{Y^2} + \frac{\tau_{xy}^2}{S_{12}^2} = 1 \tag{36}$$

Hoffman[307] modified Hill's analysis to include differences in tensile and compressive strengths and also applied this to composites. Hu[308] presented a rather limited yield criterion for anisotropy using the Tresca (maximum shear stress) criterion. The restriction arises because he required the principal stress axes to be coincident with the material's natural axes.

St. Venant's maximum strain theory was successfully applied to orthotropic composites by General Dynamics Corp., Fort Worth Division,[309] to organic matrix systems. The theory, as applied to anisotropic materials, states that failure occurs when any strain, associated with the material axes, reaches its maximum value

$$\varepsilon_x = e_1$$
$$\varepsilon_y = e_2 \tag{37}$$
$$\varepsilon_{xy} = e_{12}$$

By using the constitutive elastic relations for an orthotropic body, solving the resulting equations for σ_1, σ_2, and τ_{12}, and substituting the limiting strains, for ϵ_x, ϵ_y, and γ_{xy}, the failure envelope is derived

$$\sigma_2 = -\frac{1}{\mu_{12}}\left(e_1 E_1 - \sigma_1\right), \quad \tau_{12} = \gamma_{12} G_{12}$$

$$\sigma_2 = E_2\left(e_2 + \frac{\mu_{12}}{E}\sigma_1\right) \tag{38}$$

Equation 38 can be reduced to the same form as the maximum strain criterion for isotropic materials by using the relationships

$$K_1 = E_1 e_1$$

$$K_2 = E_2 e_2 \tag{39}$$

$$k_{12} = G_{12}\gamma_{12}$$

Substituting Equation 39 into 38 yields

$$\sigma_1 - \mu\sigma_2 = \sigma_0$$

$$\sigma_2 - \mu\sigma_1 = \sigma_0 \tag{40}$$

for isotropic materials where

$$\sigma_0 = K_1 = K_2, \quad E_1 = E_2, \quad \mu_{21} = \mu_{12} = \mu \tag{41}$$

and where σ_1 and σ_2 must be principal stresses.

In the preceding discussion all the yield criteria are only mathematical theories representing experimental data. The ultimate test for the validity of these theories is correlation with data for composite materials. The assumptions on which these theories are based may, in fact, be invalid. A few correlations to organic composites have been made but the work on metal matrix composites has been rather limited. An operationally simple strength criterion for anisotropic materials has recently been proposed by Tsai and Wu,[310] which appears to be an important improvement over existing quadratic approximations of the yield surface by using strength tensors. Here, as in most cases, the approach has been applied to a graphite/epoxy composite but not to a metal matrix composite. The work of Hecker, Hamilton, and Ebert[311] on the analysis of residual stresses and axial loading in metal matrix composites is also most interesting. It would be helpful to extend this approach to systems other than the copper core-steel case model system employed in their study.

Interpolation of the three most common theories to off-axis loading has been listed by Kreider[123] as

I. MAXIMUM STRESS THEORY

$$\sigma_0 \leq X/\cos^2\theta$$

$$\leq Y/\sin^2\theta \tag{42}$$

$$\leq S/\sin\theta\cos\theta$$

II. MAXIMUM STRAIN THEORY

$$\sigma_0 \leq X/(\cos^2\theta - \mu_{12}\sin^2\theta)$$

$$\leq Y/(\sin^2\theta - \mu_{21}\cos^2\theta) \tag{43}$$

$$\leq S/\sin\theta\cos\theta$$

III. MAXIMUM WORK THEORY

$$1/\sigma_0^2 = \cos^4\theta/X^2 + \sin^4\theta/Y^2$$

$$+ [1/S^2 - 1/X^2]\cos^2\theta\sin^2\theta \tag{44}$$

where

X = composite strength parallel to the fibers
Y = composite strength perpendicular to the fibers
S = shear strength of the composite

and the failure modes are predicated on the load orientation where

1. fiber tensile failure is expected at low angles;

2. matrix shear failure is expected at intermediated angles;

3. matrix plane strain tensile failure is expected at angles near 90°.

Since the failure theories may not be accurate, these interpolation schemes are also questionable. They are, however, the only ones available.

Fracture Mechanics

Tetleman[312] points out that, as in the case of homogeneous materials, the process of fracture of composites involves three steps: 1. the initiation of a microcrack; 2. the stable growth of this microcrack under increasing load to macrocrack size; and 3. the unstable propagation of this crack at a critical stress level.

Concepts of linear elastic fracture mechanics are the principles used to predict conditions which cause step 3 above just prior to general failure. For an isotropic, homogeneous material, the well-known Griffith-Irwin relationship defining the gross fracture stress of an infinite plate is[313]

$$\sigma_f = \left[\frac{GE}{\pi C}\right]^{1/2} \tag{45}$$

where

σ_f = gross fracture stress

C = half crack length at the onset of phase (3)

E = elastic modulus

G = crack extension force or energy per unit area of crack extension

Equation 45 is expressed in terms of the stress intensity factor K as

$$K = \sigma(\pi C)^{1/2} \qquad (46)$$

The critical stress intensity factor in the fully elastic case for an infinite plate subjected to plane stress is

$$K = (GE)^{1/2} \qquad (47)$$

Beyond this value unstable crack growth begins. Kreider[123] notes that anisotropy complicates the analysis and the inhomogeneity further complicates the solution.

Tetleman[312] attributes fracture initiation to one of four processes: 1. fiber fracture, 2. interface shear failure (debonding), 3. matrix shear fracture, or 4. matrix tensile fracture, depending upon the corresponding values of applied strain. He further divides initial microcrack growth into three categories: 1. all filaments have the same strength; 2. filaments have a distribution of strengths, where the matrix is brittle or semi-brittle; and 3. filaments have a distribution of strengths, where the matrix is relatively tough. The third category is most representative of metal matrix composites. In this case the matrix can support the stress concentrations developed by the fracture of the lower strength fibers and, by localized plastic flow, transmit the load to adjacent fibers, raising the stress in these fibers above the average of the composite.

In longitudinal crack propagation where the matrix is tough and the matrix-fiber bond is strong, fibers ahead of the advancing crack front may be broken while the remaining bridge of matrix can then neck down and fracture is completely ductile. If the fiber-matrix interface is weak, advancing cracks may be blunted by debonding perpendicular to the direction of crack propagation.

Discontinuous Reinforcement (Whiskers)

Rigorous analyses of these systems have been scarce and are not considered here. Instead,

recognizing the limitations, the shear-lag analysis will be examined briefly. This analysis is a one-dimensional model formed by imbedding a single inclusion in an infinite matrix. For continuous fibers the stress is applied to the fibers by the shearing stresses at the interface. If the normal tensile stresses in the fiber are built up linearly from the fiber ends where $\sigma_f = 0$ (good for a first approximation), by balancing forces in the fiber at the interface a critical length, ℓ_c, can be derived. This length is necessary to achieve the full strength of the fiber and, hence, cause fracture to initiate in the fiber and obtain maximum reinforcement. As developed by Kelly and Davies[174]

$$\ell_c = 2\left(\frac{r\sigma_f}{2\tau}\right) = \frac{d\sigma_f}{2\tau} \qquad (48)$$

where

r = fiber radius

σ_f = maximum strength of the reinforcement phase

d = fiber diameter

τ = shear strength of the matrix or interface depending upon which is the smaller

From Equation 48 the critical aspect ratio is defined as

$$\ell_c/d = \frac{\sigma_f}{2\tau} \qquad (49)$$

This ratio defines the minimum reinforcement geometry necessary to achieve reinforcement of the matrix.

If τ is constant then the average stress in a fiber stretched almost to the breaking strain in the central region is

$$\bar{\sigma} = \frac{\sigma_f}{\ell}(\ell - \ell_c) + \frac{\sigma_f}{\ell}\frac{\ell_c}{2} = \sigma_f\left(\ell - \frac{\ell_c}{2}\right) \qquad (50)$$

If τ is not constant, the average stress can be written

$$\bar{\sigma} = \sigma_f\left[1 - (1 - \beta)\frac{\ell_c}{\ell}\right] \qquad (51)$$

where $\beta\sigma_f$ is the average stress in a fiber within a distance $\ell_c/2$ from either end.

The strength of aligned discontinuous fibers in uniaxial tension can then be determined by substituting Equation 51 for σ_f in Equation 5[174]

$$\sigma_c = \sigma_f V_f\left[1 - \frac{1 - \beta}{\alpha}\right] + \left(\frac{d\sigma_m}{d\epsilon_m}\right)_\epsilon (1 - V_f) \qquad (52)$$

where $a = \dfrac{\ell}{\ell_c}$ and it can be shown that the theoretical strength of discontinuous reinforcements is always less than continuous fibers.

This is demonstrated in the following equation:

$$\frac{\sigma_c^{disc}}{\sigma_c^{cont}} = 1 - \frac{\alpha}{1-\beta} \frac{1}{\left[1 + \dfrac{\left(\dfrac{d\sigma_m}{d\varepsilon_m}\right)\varepsilon}{\sigma_f} \left(\dfrac{1}{V_f} - 1\right) \right]} \qquad (53)$$

Chapter 15

IMPROVED MECHANICAL PROPERTIES

While it is not possible to keep a manuscript current through the throes of preparation and assembly, it has seemed wise to reserve a section toward the end where the latest property data could be considered. We will examine, albeit briefly, a few of the most recent developments in the field. This is particularly appropriate because it has become apparent that a major advance has been achieved in the mechanical properties of metal matrix composites within the past year. This advance rests on the use of an improved, wider diameter boron filament. Some information on the newer boron filaments was presented in the section on reinforcements. The transverse strength of the wider diameter filaments was much improved, and the reliability of the longitudinal strength was greater. For example, the amount of 5.6 mil boron filaments below 400,000 psi longitudinal tensile strength was only 10% compared to 20% of the 4.0 mil boron filaments. This is perhaps more significant than the increase in average tensile strength by about 25,000 psi since it is the weak filaments that will fail first,

and perhaps initiate premature composite failure. Filament splitting has also been substantially reduced, eliminating a major problem in premature failure. As was true in the earlier discussions of mechanical behavior of metal matrix composites, consideration has been confined to the well advanced systems of B/Al and B/Ti.

The properties of B/Al composites as reported by various commercial sources have been compiled in Table 23. There are two trends that should be noted. One is that the wide diameter filaments give consistently higher composite strengths in both the longitudinal and transverse orientations. The other is that heat treatment to the T-6 condition for an aluminum alloy with wide diameter filament improves the properties significantly. More will be said on this point subsequently but the improvement in strengths of composites containing the smaller diameter filaments by heat treatment is not significant and shows no particular trend if more illustrations are given (as in Figure 110). The maximum values of 230,000 psi for longitudinal tensile strength of B/6061-T6

TABLE 23

Properties of 50 v/o Boron/Al

Source	Filament	Matrix & heat treatment condition	Longitudinal tensile strength (ksi)	Transverse (ksi) tensile strength
A	Borsic/4.2 mil	6061-F; 2024-F	165	13
A	Borsic/4.2 mil	6061-T6; 2024-T6	180	20
B	Boron/4.0 mil	6061-F	180	13.5
B	Boron/4.0 mil	6061-T6	205	20
C	Boron/4.0 mil	6061-F	175	10
C	Boron/4.0 mil	6061-T6	170	15
A	Borsic/5.7 mil	6061-F	190	19
A	Borsic/5.7 mil	2024-F	190	27
A	Borsic/5.7 mil	6061-T6	200	36
A	Borsic/5.7 mil	2024-T6	200	45
B	Boron/8.0 mil	6061-F	210	18
B	Boron/8.0 mil	6061-T6	230	37
C	Boron/5.6 mil	6061-F	180	18
C	Boron/5.6 mil	6061-T6	175	25

Al and 45,000 psi for transverse tensile strength of B/2024-T6 Al are indeed encouraging. The properties of several B/Ti composites are given in Table 24. Here, the excellent longitudinal and transverse strengths of the composite with large diameter boron are easily seen. The values of 185,000 psi and 64,000 psi, respectively, for a B/Ti-6Al-4V composite are quite promising. The transverse tensile strength of the titanium composite containing wide diameter boron is markedly superior, just as was the result for aluminum matrix composites. There is also an interesting improvement in the longitudinal tensile strength of Ti-6Al-4V alloy composites.

A critical property evaluation of the best available aluminum and titanium alloy matrix composites with boron filament reinforcement has recently been completed by Kreider, Dardi, and Prewo.[314] This investigation has produced an important body of information on composite behavior including bonding conditions, environmental effects, transverse properties, off-axis properties, failure mechanisms in fatigue, notch bending fracture, and notched tensile fracture behavior. A few of their major findings will be considered here. Longitudinal tensile strengths for 5.6 mil B with various Al alloys are shown in Table 25. A very high volume loading was achieved, as high as 70 v/o B in 2024 Al. Thus, strengths to nearly 280,000 psi were obtained with an elastic modulus value of 40×10^6 psi. Longitudinal strengths for Borsic/Al are listed in Table 26. Here, volume loadings of about 60% Borsic have been obtained with strengths exceeding 200,000 psi and moduli approaching 40×10^6 psi. The transverse tensile strengths for 5.6 mil B in various Al alloys are shown in Table 27. The transverse strength of B/2024-T6 Al in this case is almost 50,000 psi for a 45 vol % loading. The heat

treatment to the T6 condition again much improves the values. This importance of T6 treatment is seen in Table 28, where an extensive series of tests is reported for 5.7 mil Borsic/Al composites. In many cases the transverse strengths were doubled.

The change in longitudinal tensile strength with temperature for a 5.6 mil B/6061 Al composite is given in Table 29. The decrease in strength with temperature is similar to that found for 4.0 mil boron composites. The transverse strength curve with temperature is much higher for the wide diameter filament reinforcement at low temperatures in a heat treated matrix than for the smaller diameter filament or an untreated matrix. At temperatures of about 300° to 400°F (149° to 204°C) the values start to converge as the wide diameter filament is not as determinant as the loss in matrix strength. By 600°F (315°C) there is little difference in any of the results. At the intermediate temperatures, however, the advantage of wide diameter filament on the transverse properties could be significant.

The off-axis behavior of 50 v/o 4.2 mil Borsic/6061F Al and 63 v/o 5.7 mil Borsic/6061-T6 Al was also investigated by Kreider, Dardi, and Prewo.[314] Specimens of 10 layer unidirectional reinforced composites, 6 in. long and about 0.5 in. wide, were tested in a series of filament orientations in an Instron Tensile Machine. For the 4.2 mil Borsic material the results were $E_{11} = 33.8 \times 10^6$ psi, $E_{22} = 20.5 \times 10^6$ psi, $G_{12} = 8.15 \times 10^6$ psi, and $V = 0.24$. For the 5.7 Borsic material the results were $E_{11} = 43.7 \times 10^6$ psi, $E_{22} = 28.3 \times 10^6$ psi, $G_{12} = 11.1 \times 10^6$ psi, and $V_{12} = 0.24$. The results of the tests for one of the composite systems are given in Table 30.

The longitudinal tensile strengths for various

TABLE 24

Properties of 50 v/o Borsic/TI

Source	Filament diameter	matrix	Longitudinal tensile strength (ksi)	Transverse (ksi) tensile strength
A	4.2 mil	Beta III	170	30
A	4.2 mil	6Al-4V	155	35
B	4.2 mil	6Al-4V	140	42
C	5.7 mil	6Al-4V	185	64

TABLE 25

Axial Tensile Strength of 5.6 mil B/Al

Matrix	v/o Boron (%)	Ultimate tensile strength (10^3 psi)	Elastic modulus (10^6 psi)	Strain to fracture (%)
2024F	45	185.7	30.4	0.765
	45	197.5	27.5	0.835
	44	177.0	30.0	0.725
	47	212.0	32.0	0.825
	47	212.0	32.6	0.820
	49	194.0	32.0	0.740
2024-T6	46	202.5	32.8	0.75
	46	213.6	31.6	0.81
	47	217.0	32.3	0.830
	48	213.0	31.3	0.845
	64	279.0	40.0	0.755
2024F	70	279.5	–	–
	66	253.0	–	–
	67	250.2	–	–
6061F	48	196.3	31.8	0.710
	48	171.0	28.2	0.590
	50	204.0	33.8	0.72
	50	208.0	32.0	0.76
6061-T6	52	216.5	33.8	0.78
	51	197.0	33.4	0.69
	50	203.0	–	–

TABLE 26

Axial Tensile Strength of 5.7 mil Borsic/Al Composites

Matrix	v/o BORSIC (%)	Ultimate tensile strength (10^3 psi)	Elastic modulus (10^6 psi)	Strain to fracture (%)
6061-F	30	115.0 113.3	17.6 18.9	0.71 0.71
6061-T6	30	156.2 152.4		
6061-F	54	203.4 181.5 199.0	36.6 36.0 36.1	0.675 0.630 0.655
6061-F	56	214.0 212.0		
6061-F	57	228.0 222.0		
6061-F	58	227.0 219.0 216.0 222.0		
6061-F	61	199.0 207.6	39.4	0.57
2024-F	58	211.5 221.0		
2024-T6	61	235.0 210.0		
5052/56	59	177.6 182.0	37.7	0.54
1100/1145	57	158.2 175.5		

volume fractions of the wide diameter boron filament in Ti-6Al-4V are given in Table 31.[314] Various bonding temperatures were employed. In Table 32 the results of testing for longitudinal strengths of 45 v/o 5.7 mil Borsic/Ti-6Al-4V by Toth[315] are given. The bonding temperature in this case was 1500°F (815°C), so some reservations must be made in comparing the two series of results. Both indicate that promising longitudinal and transverse strengths can be developed in B/Ti composites utilizing appropriate bonding conditions.

The marked gain in transverse strengths utilizing a wider diameter boron is indicated in Figure 153. The composite transverse strength has been plotted as a function of the matrix tensile

TABLE 27

Transverse Tensile Properties of 5.6 mil B/Al

Matrix	v/o Boron	UTS (10^3 psi)	Elastic modulus (10^6 psi)	Strain to failure (%)
2024F	45	27.0		
	45	27.2		
	45	26.2		
2024-T6	45	48.0		
	45	48.7		
	45	38.2		
2024-T6	55	41.9	21.0	0.23
		45.0	22.5	0.24
2024F	66	26.0		
		27.3		
6061F	50	18.9		
	50	19.0		
	50	18.3		
6061-T6	50	34.8		
	50	37.4		
	50	41.7		

strength. Previous data with 4.2 mil Borsic gave composite strengths relatively independent of the matrix strength. In other words, it didn't do any good to heat treat the matrix to higher strengths. This could be attributed to the premature failure of the composite caused by filament splitting. Thus, the filament sufficiently embrittled or strained the matrix so that when it failed, the composite failed. With the higher quality filament, the strength of the matrix now plays an important role in determining the composite transverse strength. Heat treatments to higher matrix strengths now have a marked effect on increasing strengths for B/Al. The shaded area indicates the results for earlier work with smaller diameter boron.

The foregoing results show the importance of generating engineering data on the composite structures with wide diameter boron. Unfortunately, most current studies[51,316] are being conducted on 4 mil filament and the design data

which may result are not as attractive as is otherwise the case. Some manufacturers are reporting the results with wide diameter filament and confirming the superiority of composite properties seen in the research efforts.[50]

In a recent large-scale production - fabrication study for aircraft structures,[316] complicated structural shapes such as tapered frames and hat sections were successfully fabricated by tape consolidation methods for B/Al composites. It was demonstrated that metal matrix aircraft structures could be fabricated at reasonable costs using conventional metal working methods. The only serious problem encountered with low transverse strength of plasma spray composites could have been eliminated by using a wider boron filament. Therefore, considerable emphasis should be made on generating data on the improved composite materials before further fabrication studies are undertaken.

TABLE 28

Transverse Tensile Properties of 5.7 mil Borsic/Al

Matrix	Boron v/o	UTS (10^3 psi)	Elastic modulus (10^6 psi)	Strain to failure (%)
6061-F	30	16.0 14.0	15.0	0.45
6061-T6	30	35.1 34.6	15.5	0.31
6061-F	52	19.3 20.5 19.3		
6061-T6	52	31.2 36.6 39.8	18.9 19.0 21.0	0.19 0.23 0.24
6061-F	60	19.9 19.7	24.0	0.26
6061-T6	60	44.0 37.2		0.23
6061-F	70	16.8 21.2 18.8	32.3	0.29
6061-T6	70	35.4 27.6	29.2	0.14
2024-F	56	22.4 20.8	22.8	0.171
2024-T6	56	46.0 37.1	24.7	0.202
5052/56	57	30.4 32.6	25.8	0.308 0.50
1100	54	13.6 13.8 11.7	21.8 23.2	0.68 0.52 0.60

TABLE 29

Axial Tensile Strength of 48 v/o 5.6 mil B/6061 Al Composites

Test temperature (°F)	UTS (10^3 psi)
70	196.3
70	171.0
500	177.0
500	135.2
900	143.8
900	137.3

TABLE 30

Results of 5.7 mil Borsic/6061-T6Al Off-Axis Tests

Specimen number	Fiber orient (degrees)		UTS (ksi)	S_{11} (10^{-8} in²/lb)	S_{12} (10^{-8} in²/lb)	S_{16} (10^{-8} in²/lb)
1	0		214	2.31	−0.54	−0.06
2**	0		192	–	–	–
3	0		208	2.28	−0.56	−0.05
		Avg.:	205	2.29	−0.55	−0.05
		Calc*:	205	2.29	−0.55	0.0
4	10		113	2.43	−0.57	0.82
5	10		134	2.29	−0.59	0.92
6	10		129	2.75	−0.69	0.60
		Avg.:	128	2.49	−0.62	0.78
		Calc.:	111	2.39	−0.61	0.55
7**	30		40.1	3.40	–	–
8	30		48.4	2.87	−0.90	1.14
9**	30		51.3	–	–	–
		Avg.:	49.4	3.13	−0.90	1.14
		Calc.:	47.2	2.99	−0.94	0.99
10	45		38.9	3.57	−1.18	0.74
11	45		39.1	3.36	−1.12	0.83
12	45		39.2	3.66	−1.15	0.49
		Avg.:	39.1	3.53	−1.15	0.69
		Calc.:	36.6	3.43	−1.07	0.62
13	90		34.9	3.50	−0.54	−0.15
14	90		31.2	3.53	−0.58	−0.08
15	90		30.5	3.55	−0.58	−0.01
		Avg.:	32.2	3.53	−0.57	−0.08
		Calc.:	32.2	3.53	−0.57	0.0

*UTS calculated using maximum work theory with X = 20.5 ksi, Y = 32.2 ksi, S = 22.2 ksi; compliance coefficients calculated using the transformation equations with E_{11} = 43.7 x 10^6 psi, E_{22} = 28.3 x 10^6 psi, G_{12} = 11.1 x 10^6 psi, v_{12} = 0.24.

**Strain gage failure.

TABLE 31
5.7 mil Borsic/Ti Composites

Matrix alloy	Volume fraction fiber %	Orientation	Bonding temperature	Elastic modulus psi	Ultimate tensile strength psi	Failure strain %
Ti6/4	0	–	As rcvd.		160,000	8.0
Ti6/4	0	–	800°C		166,000	7.0
Ti6/4	23	0°	800°C		174,000	
					191,000	
Ti6/4	23	90°	800°C		47,500	
					48,300	
Ti6/4	45	0°	750°C		133,000	
					133,000	
Ti6/4	43	0°	870°C	33x10^6	180,000	0.52
					186,000	
					176,000	
Ti6/4	40	90°	870°C	29x10^6	54,000	0.46
					56,000	
				31x10^6	48,000	0.28
Ti6/4	40	90°	870°C	27x10^6	54,000	0.42
				28x10^6	49,500	0.34
Tested at 600°F		90°	870°C		42,000	
					41,200	
Tested at 1000°F		90°	870°C		33,200	
					24,200	

TABLE 32
Elevated Temperature Tensile Tests
45 v/o 5.7 mil Borsic/Ti-6Al-4V

Specimen no.	Test orientation	Test temp. (°F)	UTS psi
1	0	R.T.	178,900
2	0	R.T.	181,800
1	90	R.T.	64,700
2	90	R.T.	63,500
3	0	500	150,500
4	0	500	155,000
3	90	500	49,200
4	90	500	48,300
5	0	800	145,200
6	0	800	140,000
5	90	800	43,700
7	90	800	46,400
8	0	1000	124,600
9	0	1000	120,200
9	90	1000	40,200
10	90	1000	39,600

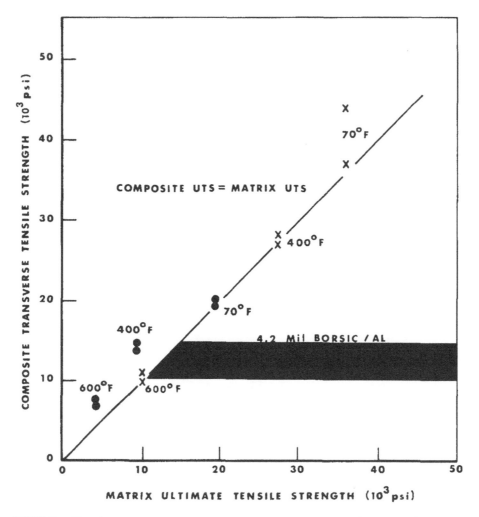

FIGURE 153. Composite transverse as a function of matrix UTS for 60 v/o-5.7 mil Borsic-6061.

Chapter 16

SOME PROSPECTS FOR THE FUTURE

Up to this point we have placed much emphasis on systems where the largest amount of property data and behavior characteristics has been acquired. This has for the most part dealt with metal matrix composites of interest to the aerospace community. In trying to assess the future technological impact of these new materials, however, it is prudent to look beyond the rather narrow confines of the aerospace propulsion and structural interests and concepts for a larger view of potential applications. Mention has been made, for example, of the possible uses of eutectic composites in electrical devices. The InSb-NiSb system was mentioned as one in which needles of NiSb of high electrical conductivity act as finely dispersed metallic inclusions in a semiconducting matrix of InSb which permits large thermomagnetic and galvanomagnetic effects. In an aligned eutectic of InSb-NiSb, one has a magnetoresistive device with important applications in a field plate (attached to an insulating surface substrate as a variable shaped 25 μm thick eutectic layer). These applications according to Weiss[203] include

1. measurement of magnetic fields and field-dependent quantities
2. contactless variable resistance
3. modulation of small dc currents and voltages
4. multiplication
5. contactless control

An example of one of these devices in use is a contactless nominal current indicator for the brake of an electric locomotive. Because of the strong absorption characteristics of NiSb needles in the infrared region of 9 to 15 μm, where InSb does not appreciably absorb, the heat difference and subsequent thermoelectric voltage from the NiSb absorption of energy have been utilized to develop an InSb-NiSb far infrared detector.[203]

Galasso[24] lists a number of possible nonstructural applications of eutectic composites, including a Cr-Cr_2O_3 eutectic which contains highly conductive Cr rods in a Cr_2O_3 insulating matrix. The Al-Al_3Ni system is suggested as a possibility for high strength conducting wire. Refractory metal wire reinforcement of copper such as in the tungsten reinforced copper composites has been seen to offer practical high strength electrical conductors from metal matrix composites.[147,148]

In their book, Davis and Bradstreet[34] have devoted a chapter to the consideration of future applications and markets for metal matrix composites. They have provided an excellent background of potential areas for use of metal composite structures where the advantages of material performance may outweigh the higher initial costs compared to most monolithic materials. In such a context, the materials become cost-effective and will likely play an expanding role in materials technology. Areas of forecasted composite usage include aircraft and aerospace, engines and turbines, pumps and compressors, railroad equipment, building and bridge construction, and ship building. One example given by Davis and Bradstreet[34] is that of the Verrazano Narrows Bridge in New York with a 4,260 ft long span and 160,000 tons of structural steel. With a B/Al composite structure they estimate that an 8,000 ft long span could have been built weighing 93,000 tons. The reader can supply his own examples without trouble, so only one more illustration will be given. High speed railway transportation would appear to be a potential candidate for composite structures because of the high stresses on structural members. A much lighter weight train would also greatly simplify roadbed requirements.

Not much has been said of competitive resin composite materials thus far. There are advantages in each type composite peculiar to the system. Resin composites are mainly filaments glued together by a small amount of weak matrix. Metal composites are more likely to be close to equivolume combinations of filament and matrix. The metal matrix is stronger, has better properties in more cases, and provides high temperature stability. The resin composite has a wealth of prior fabrication techniques from fiberglass technology and is substantially more economical to make. The authors feel that as composite technology progresses, these composite system approaches will complement each other more than compete. For lower temperatures the resin-based composites will likely be more cost-effective, where at higher temperatures, or in adverse environments, the metal-based composites may well be more attractive.

For specific demonstration hardware studies of metal matrix composites it is necessary to return to the area of prime current technical interest, in aerospace structures. The first significant effort was to fabricate a B/Al prototype of a missile payload adapter which was 60 in. in diameter and 42 in. in height.[177,179] This structure was 40% lighter made out of B/6061 Al than the original Al-steel structure. This was the first demonstration of real weight savings in an aerospace structure.

The feasibility of using metal matrix composites has been examined in other space structures. A panel of the Apollo Service Module was selected for one study on B/Al composites.[317] In this case, the lack of a high quality composite sheet material greatly impeded the program.

In another effort a typical aircraft wing spar beam was fabricated from B/Al composite material.[318] Recently, the range of usefulness of metal composite materials has been expanded considerably for primary aircraft structures such as an aircraft bulkhead, using B/Al.[51] Substantial weight savings were realized. A number of complex shapes have been successfully fabricated for aircraft structures.[316] Initial attempts to fabricate Borsic/Al propeller blades were also successful.[319] These have been followed by a considerable number of probing efforts on B/Al and Borsic/Ti for turbine blades. In one such program gas turbine compressor blades were fabricated and tested under simulated service conditions.[320] Except for a slight erosion problem, the composite blades outperformed monolithic titanium blades tested under the same conditions. A first stage fan assembly with B/Al composite blades is shown in Figure 154. It is possible that an advanced turbine engine design

FIGURE 154.   B/Al compressor fan blades.

incorporating B/Al or Borsic/Al compressor blades will be an early major application of metal matrix composites in aircraft. Weight savings of 30 to 40% compared to titanium alloy blades have been reported.

Widespread acceptance of metal composite structures is not likely unless the prices continue to decline. Fortunately, this has been the case for B/Al as shown in a cost history projection by Alexander,[321] in Figure 155. For comparison the cost analysis of Toth[322] is given in Figure 156. These figures can be contrasted to the cost for continuous alumina filament that remains at $7,500 to $10,000/lb or more, and the price of alumina whiskers which exceeded $5,000/lb after over a decade of development. The need to develop cheaper filaments of reasonable quality is most pressing. For boron the picture improves even more with the acceptance of a larger diameter filament. The 1971 average price for 4 mil boron filament was about $210/lb. With 5.6 mil boron a cost reduction of almost 50% is obtained. Further reductions with increasing volume are expected. The cost of a boron aluminum monotape is expected to be as low as $75/lb by 1976 according to Toth.[322] Given the high property levels for advanced composite materials with wide diameter boron, at this point they would provide real competition to some aerospace alloys. By 1980 it is possible to project costs of $50/lb or less, assuming a reasonable market exists so that quantity production is necessary. These projections involve considerable guesswork, but at least the general trend for B and B/Al has been steadily downward in price.

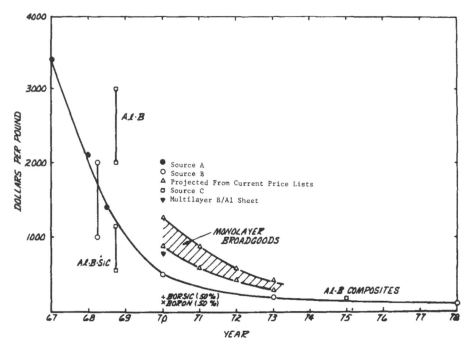

FIGURE 155.   Metal matrix composite cost history and projections.

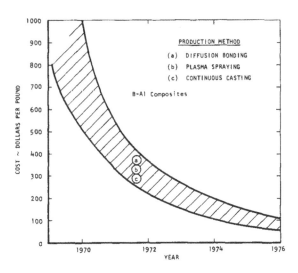

FIGURE 156.   Current and projected costs for fibers and composites.

Chapter 17

CONCLUSIONS

Sufficient mechanical property data have been obtained on several metal· matrix composite systems to demonstrate that efficient reinforcement can be achieved. The high strength, high modulus continuous filaments showing the most promise for metals have been boron and Borsic, with alumina attractive for its stability at elevated temperatures toward many transition metals such as titanium and nickel.

Both boron/aluminum and boron/titanium have been fabricated with encouraging mechanical properties, particularly in the direction of the reinforcement. These systems have been extensively studied. The controlled interaction between the matrix and reinforcements during fabrication has been achieved through diffusion-barrier coatings such as silicon carbide and boron nitride on boron and through use of alloy additions to decrease matrix-filament reactivity. Fabrication parameters have been optimized to minimize filament degradation while still achieving an adequate interfacial bond.

The longitudinal tensile and compressive strengths, creep resistance, and fatigue behavior of metal matrix composites have been excellent. The mechanical properties are quite sensitive to filament-load angle. The transverse properties of unidirectional composites have been low, evidently due to factors such as fiber splitting. Methods of improving transverse strength include development of improved filaments, matrix heat treatment, and use of cross or angle plies (with attendant loss in structural efficiency). The recent use of wider diameter boron filaments in boron/aluminum and boron/titanium composites has resulted in a major advance in mechanical properties, and particularly the significant improvement in transverse strengths. There is a need to obtain adequate engineering data on these latest, improved composite structures.

For very high temperature applications, two distinct approaches have emerged. One system uses the reinforcement of superalloys with refractory metal wires. The other system consists of directionally solidified eutectic composites. Both approaches show considerable potential, but need substantial further research to produce practical structures.

Whisker reinforced composites exhibit some improvement in mechanical properties over the metals themselves but do not indicate a high degree of efficiency in utilizing the exceptional tensile strengths of the whiskers. The high temperature properties deteriorate rapidly where processing causes substantial filament breakage.

Methods are available for the prediction of the mechanical properties of composite structures and their response to external loads. Most of these methods, however, have yet to be applied to the prediction of properties of metal matrix composites and the subsequent comparison to experimental results.

Much of the early emphasis on metal matrix composites has come from the aerospace industry. Feasibility studies have clearly shown significant weight savings and improved performance in a number of aerospace applications. As a wide variety of consolidation and fabrication processes are developed, properties become more reliable, and the unit costs of filament, monolayer tape, and composite sheet material are reduced, other nonaerospace applications should become of increasing interest. Possibilities for the future would include such applications as bridge construction and high speed railway trains. Very important nonstructural uses are likely in reinforced cables and electrical devices.

REFERENCES

1. Sinclair, P. M., *Ind. Res.,* 11, 58, 1969.
2. Alexander, J. A., Shaver, R. G., and Withers, J. C., Critical analysis of accumulated experimental data on filament reinforced metal matrix composites, NASA Rep. NASW-1779, June 1969.
3. Rauch, H. W., Sr., Sutton, W. H., and McCreight, L. R., *Ceramic Fibers and Fibrous Composite Materials,* Academic Press, New York, 1968, 395.
4. Burte, H. M. and Lynch, C. T., The technical potential for metal matrix composites, *Metal Matrix Composites,* ASTM STP-438, 3-25.
5. Burte, H. M. and Lynch, C. T., Defense Metals Information Center Memorandum 243, May 1 to 6, 1969.
6. Tsai, S. W., Some problems related to fiber-reinforced metals, presented at the Metal Matrix Working Group Meeting, USAF Academy, Colorado, January 13, 1970.
7. Anon., Boron filament process development, Boron Trichloride-Tungsten Process, Vol. 1, AFML-TR-67-120, May 1967.
8. Snide, J. A., Lynch, C. T., and Whipple, L. D., Current developments in fiber-reinforced composites, AFML-TR-67-359, February 1968.
9. Lasday, A. H. and Talley, C. P., Boron filament for structural composites, in Advanced Fibrous Reinforced Composites, 10th SAMPE Symp., San Diego, Calif., November 9 to 11, 1966.
10. Galasso, F. S., *J. Appl. Phys.,* 38 (1), 414, 1967.
11. Burte, H. M., Bonanno, F. R., and Herzog, J. A., Metal matrix composite materials, in Orientation Effects in the Mechanical Behavior of Anisotropic Structural Materials, ASTM STP-405, American Society for Testing and Materials, 1966, 59.
12. Galasso, F. S. and Paton, A., *Trans. AIME,* 236 (12), 1751, 1966.
13. Reeves, R. B. and Gebhardt, J. J., Preparation of boron filament on fused silica substrates by decomposition of diborane, in Advanced Fibrous Reinforced Composites, 10th SAMPE Symp., San Diego, Calif., November 9 to 11, 1966.
14. Basche, M., Fanti, R., and Galasso, F. S., *Fiber Sci. Technol.,* 1 (1), 19, 1968.
15. Vidoz, A. E., Camahort, J. L., and Crossman, F. W., *J. Comp. Mater.,* 3, 254, 1969.
16. Withers, J. C., Handlen, L. C., and Schwartz, R. T., Continuous silicon carbide filaments, in Advanced Fibrous Reinforced Composites, 10th SAMPE Symp., San Diego, Calif., November 9 to 11, 1966.
17. LaBelle, H. E. and Hurley, G. F., *SAMPE J.,* 6 (1), 17, 1970.
18. Lyon, S. R., private communication, Air Force Materials Laboratory, January 1970.
19. Dauksys, R. J. and Ray, J. D., *J. Comp. Mater.,* 3, 684, 1969.
20. Jackson, P. W., Some Studies of the Compatibility of Graphite and Other Fibers with Metal Matrices, WESTEC Conf. 1969, ASM, Los Angeles, March 10 to 13, 1969.
21. Ray, J., Gloor, W., and Ross, J., private communication, Air Force Materials Laboratory, March 1970.
22. Moore, T. L. and Snide, J. A., High temperature compatibility of AFC-77 and titanium, AFML-TR-68-110, July 1968.
23. Petrasek, D. W. and Signorelli, R. A., Preliminary evaluation of tungsten alloy fiber-nickel alloy composites for turbojet engine applications, NASA TN D-5575, Washington, D.C., February 1970.
24. Galasso, F. S., *High Modulus Fibers and Composites,* Gordon and Breach Science Publishers, New York, N.Y., 1969.
25. Branch, K., private communications, University of Dayton Research Institute, 1968-1969.
26. Crane, R. L. and Tressler, R. E., *J. Comp. Mater.,* 5, 537, 1971.
27. Tressler, R. E. and Crane, R. L., Sapphire filament mechanical property consideration of importance to Al_2O_3 reinforced metals, Am. Ceram. Soc., 73rd Ann. Meeting, Chicago, Ill., April 27, 1971.
28. Mellinder, F. P. and Proctor, B. A., *Phil Mag.,* 13, 197, 1966.
29. Crane, R. L. and Tressler, R. E., Application of phase equilibria and reactivity studies in the system Ti-Al-O to $Ti-Al_2O_3$ composite stability, Am. Ceram. Soc., 73rd Ann. Meeting, Chicago, Ill., April 27, 1971.
30. Shahinian, P., *J. Am. Ceram. Soc.,* 54, 67, 1971.
31. Gruber, R. J., Continuous oxide filament synthesis (CVD), Tech. report, AFML-TR-71-251, January 1972.
32. Simpson, F. H., Continuous oxide filament synthesis (devitrification), The Boeing Company, AFML-TR-71-135, October 1971.
33. Hamilton Standard, Division of United Aircraft Corporation, Windsor Locks, Connecticut, Composite Materials Department.
34. Davis, L. W. and Bradstreet, S. W., *Metal and Ceramic Matrix Composites,* Cahners Publishing Co., Inc., Boston, Mass., 1970.
35. Kreider, K. G. and Leverant, G. R., Boron fiber metal matrix composites by plasma spraying, AFML-TR-66-219, July 1966.
36. Kreider, K. B., Schile, R. D., Breinan, E. M., and Marciano, M., Plasma sprayed metal matrix fiber reinforced composites, AFML-TR-68-119, July 1968.
37. Breinan, E. M. and Kreider, K. G., *Metals Engin. Quart.,* November 1969, 5.
38. Kreider, K. G. and Marciano, M., *Trans. Met. Soc. AIME,* 245, 1279, 1969.

39. Breinan, E. M. and Kreider, K. G., *Metallurgical Trans.*, 1, 93, 1970.
40. Tsareff, T. C., Sippel, G. R., and Herman, M., Metal matrix composites, DMIC Memorandum 243, 38, May 1969.
41. Hanby, K. R., DMIC Report S-21 (1967), June 1, 1968.
42. Rauch, H. W., Sutton, W. H., and McCreight, L. R., Survey of Ceramic Fibrous Composite Materials, AFML-TR-66-365, October 1966.
43. Cunningham, A. L. and Alexander, J. A., Advances in structural composites, SAMPE, Western Periodicals G., North Hollywood, Calif., 12, AC-15, 1967.
44. Alexander, J. A., Withers, J. C., and Macklin, B. A., Investigation of three classes of composite materials for space vehicle application, NASA CR-785, May 1967.
45. Alexander, J. A., Cunningham, A. L., and Chang, K. C., Investigation to produce metal matrix composites with high modulus, low density continuous filament reinforcements, AFML-TR-67-391, February 1968.
46. Schmitz, G. K. and Metcalfe, A. G., Development of continuous filament reinforced metal tape, AFML-TR-68-41, February 1968.
47. Wolff, E. G. and Hill, R. J., Research on boron filament/metal matrix composite materials, AFML-TR-67-140, June 1967.
48. Jackson, P. W., Baker, A. A., Cratchley, D., and Walker, P. J., *Powder Metallurgy,* 2 (21), 1, 1968.
49. Martin, G. and Moore, J. F., Research and development of non-destructive testing techniques for composites, AFML-TR-66-270, April 1966.
50. Dolowy, J. F., Jr., Mechanical properties from 5.6 mil B aluminum-boron composites, 24th Pacific Coast Regional Meeting, Am. Ceram. Soc., Anaheim, Calif., November 1, 1971.
51. Maikish, C. R., Wennhold, W. F., and Weisinger, M. D., Application of advanced metal matrix composites to aircraft structures, 24th Pacific Coast Regional Meeting, Am. Ceram. Soc., Anaheim, Calif., November 1, 1971.
52. Greening, T. A., Advanced wire-wound tungsten nozzles, AFAPL-TR-67-181, April 1967.
53. Alexander, J. A., Shaver, R. G., and Withers, J. C., A study of low density high strength, high modulus filaments and composites, NASA CR-523, July 1966.
54. Ahmad, I., Creco, V. P., and Barranco, J. M., Electroforming of the composites of nickel reinforced with some high-strength filaments, Watervliet Arsenal Tech. Rep. WVI-6709, January 1967.
55. Adsit, N. R., Fiber strengthened metallic composites, ASTM STP-427, October 1967, 27.
56. Wither, J. C. and Abrams, E. F., The electroforming of composites, *Plating*, 605, 1968.
57. Baker, A. A., Harris, J. T., and Holms, E., Fiber reinforcement of metals and a filament winding and electroforming technique, *Metals and Materials*, 2, 211, 1967.
58. Donovan, P. D. and Watson-Adams, B. R., Formation of composite materials by electrodeposition, *Metals and Materials*, 3, 443, 1969.
59. Petrasek, D. W., Elevated-temperature tensile properties of alloyed tungsten fiber composites, NASA TN-D-3073.
60. Schuerch, H., Compressive strength of boron-metal composites, NASA CR-202, April 1965.
61. Alexander, J. A. and Davies, L. G., Continuous casting as a composite fabrication process, 15th National SAMPE, Western Periodicals, North Hollywood, Calif., Vol. 15, April 1969.
62. Davies, L. G., Shaver, R. G., and Withers, J. C., Continuous cast boron-light metal alloy preforms and composites, General Technologies Corp., Reston, Va., presented at 17th Refractory Composites Working Group Meeting, June 16 to 18, 1970.
63. Hanby, K. R., Fiber reinforced metal matrix composites, DMIC Report S-33, 1969-1970, p. 37-8, July 1, 1971.
64. Snajdr, E. A. and Williford, J. F., Jr., Investigation of fiber-reinforced metal matrix composites using a high energy rate forming method, final report AFML-TR-68-252, December 1968.
65. Williford, J. F., Jr. and Snajdr, E. A., Investigation of fiber-reinforced metal matrix composites using a high energy rate forming method, Metal Matrix Composites, DMIC Memorandum 243, Battelle Memorial Inst., Richland, Washington, May 1969.
66. Robinson, R. K., High energy-rate forming of fibrous composites, Fiber-strengthened metallic composites, ASTM STP-427, Am. Soc. Testing Mats., 1967, 107.
67. Fleck, J. N. and Leonard, R. W., Explosive fabrication of metal matrix composites, Metal Matrix Composites, DMIC Memo 243, Battelle Memorial Inst., Columbus, Ohio, May 1969, 24.
68. Fleck, J. N. and Goldstein, M., Beryllium reinforced aluminum, presented at the 15th National SAMPE Symposium, Los Angeles, Calif., April 1969.
69. Metcalfe, A. G., *J. Comp. Mater.,* 1, 356, 1967.
70. Klein, M. J., Reid, M. L., and Metcalfe, A. G., Compatibility studies for viable titanium matrix composites, AFML-TR-69-242, October 1969.
71. Metcalfe, A. G. and Schmitz, G. K., Development of filament reinforced titanium alloys, presented at SAE, Aeronautic and Space Engineering and Manufacturing Meeting, Los Angeles, Calif., October 2 to 6, 1967.
72. Divecha, A. P. and Pignone, E. H., Development of a method for fabricating metallic matrix composite shapes by a continuous mechanical process, Quarterly Progress Report, NASA Contract NAS8-27010, July 1971.
73. Withers, J. C., The fabricatility of multilayer beryllium wire reinforced aluminum composites by chemical vapor deposition, GTC Tech. Report 155, 5-1, July 1968.

74. Greszczuk, L. B., Filament reinforced refractory metals, contained in *Advanced Fibrous Reinforced Composites*, Society of Aerospace Materials and Process Engineers, Western Periodicals, North Hollywood, Calif., Vol. 10, November 1966, F-15.

75. Baskey, R. H., Fiber-reinforced metallic composite materials, AFML-TR-67-196, September 1967.

76. Meyerer, W. J., The feasibility of forming a boron fiber-reinforced aluminum composite by a hot extrusion process, AFML-TR-68-127, August 1968.

77. Ebert, L. J. and Gadd, J. D., A mathematical model for mechanical behavior of interfaces in composite materials, *Fiber Composite Materials*, American Society for Metals, 1965, 89.

78. Ebert, L. J., Hamilton, C. H., and Hecker, S. S., Development of design criteria for composite materials, AFML-TR-67-95, April 1967.

79. Stuhrke, W. F., The mechanical behavior of aluminum-boron composite material, in *Metal Matrix Composites*, ASTM STP-438, American Society for Testing and Materials, 1958, 108.

80. Blackburn, L. D., Burte, H. M., and Bonanno, F. R., Filament-matrix interactions in metal matrix composites, in *Strengthening Mechanisms-Metals and Ceramics*, University Press, Syracuse, 1966.

81. Snide, J. A., Compatibility of vapor deposited B, SiC, and TiB_2 filaments with several titanium matrices, AFML-TR-67-354, February 1968.

82. Chang, Y. A., Phase compatibility studies on nickel-chromium-silicon-carbon based alloys, AFML-TR-68-63, Vol. I, May 1968.

83. Brukl, C. E., Phase compatibility studies on nickel-chromium-silicon-carbon base alloys, AFML-TR-68-63, Vol. II, January 1969.

84. Ashdown, F. A., Compatibility study of SiC filaments in commercial-purity titanium, Air Force Institute of Technology, Thesis GSF/MC/68-1, July 1968.

85. Blackburn, L. D., Herzog, J. A., Meyerer, W. J., Snide, J. A., Stuhrke, W. F., and Brisbane, A. W., Progress and planning report for MAMS internal research on metal matrix composites, MAM-TM-66-3, January 1966.

86. Forest, J. D. and Christian, J. L., Development and application of aluminum-boron composite materials, AIAA 5th Ann. Meeting, Philadelphia, October 21 to 24, 1968.

87. Hamilton, C. H., Development of Ti-6Al-4V borsic composites, Defense Metals Information Center Memorandum 243, May 1969, 33.

88. Hamilton, C. H., private communication, North American Rockwell Corp., Los Angeles, April 1970.

89. Schmitz, G. K., Klein, M. J., Reid, M. L., and Metcalfe, A. G., Compatibility studies for viable titanium matrix composites, AFML-TR-70-237, Solar Division of International Harvester Company, September 1970.

90. Wolff, E., Boron filament, metal matrix composite materials, Report No. 2 on Contract AF33(615)-3164, June 1966.

91. Klein, J. and Metcalfe, A. G., Effect of interfaces in metal matrix composites on mechanical properties, AFML-TR-71-189, Solar Division, October 1971.

92. Moore, T. L., Elevated temperature compatibility of Ni/Al_2O_3 composites, Defense Metals Information Center Memorandum 243, May 1969, 47.

93. Mehan, R. L., Stability of oxides in metal or metal alloy matrices, General Electric Co., Technical Management Report No. F33615-69-C-1635, December 1, 1969.

94. Tressler, R. E. and Moore, T. L., SEM study of sapphire filament reinforced metals, presented at the 3rd Ann. Scanning Electron Microscopy Symp., Chicago, April 28, 1970.

95. Tressler, R. E. and Moore, T. L., Mechanical property and interface reaction studies of titanium-alumina composites, presented at the 1970 WESTEC Conf. of the American Society for Metals, Los Angeles, March 10, 1970.

96. Tressler, R. E. and Moore, T. L., Mechanical property and interface reaction studies of titanium-alumina composites, *Metals Engin. Quart.*, 11 (1), 16, 1971.

97. Crane, R. L., private communications, AFML, January 1972.

98. Davison, J., personal communication, University of Dayton, 1967, Data compiled with ΔF values from Bureau of Mines Bulletin 542.

99. Sutton, W. H., Whisker composite materials — a prospectus for the aerospace designer, *Astronautics and Aeronautics*, August 1966, 46.

100. Mehan, R. L. and Harris, T. A., Stability of oxides in metal or metal alloy matrices, AFML-TR-71-150, August 1971.

101. Sutton, W. H. and Feingold, E., Role of interfacially active metals in the apparent adherence of nickel to sapphire, Mat. Sci. Res., 3, 577, 1966.

102. Chen, P. E. and Lin, J. S., Transverse properties of fibrous composites, Mat. Res. Std., 2, 29, 1969.

103. Cooper, G. A. and Kelly, A., Role of the interface in the fracture of fiber-composite materials, Chapter in Interfaces in Composites, ASTM STP-452, 1969, 90.

104. Nicholas, M., The strength of metal/alumina interfaces, *J. Mater. Sci.*, 3, 571, 1968.

105. Calow, C. A. and Porter, I. T., *J. Mater. Sci.*, 6, 156, 1971.

106. Calow, C. A., Bayer, P. D., and Porter, I. T., *J. Mater. Sci.*, 6, 150, 1971.

107. Meiser, M. and Davison, J. E., Graphite filament reinforcement of an iron-aluminum alloy, presented at the 1969 WESTEC Conference, ASM, Los Angeles, Calif., March 12, 1969.

108. Hough, R. L., Refractory diffusion barriers for SiC and B filaments in a metal matrix, AFML-TR-68-381, December 1968.

109. Hough, R. L., Development of manufacturing process for large-diameter carbon- base monofilaments by chemical vapor deposition, NASA CR-72770, November 1970.

110. Ezekiel, H. M., High strength high modulus graphite fibers, AFML-TR-70-100, January 1971.

111. Quackenbush, N. E., Large diameter graphite, carbon composite filament development, NASA CR 72769, July 1970.

112. Volk, H. F., Nara, H. R., and Hanley, D. P., Integrated research on carbon composite materials, AFML-TR-66-310, Part V, Vol. I, January 1971.

113. Pepper R. T., Rossi, R. C., Upp, J. W., and Riley, W. C., Development of an aluminum-graphite composite, Aerospace Report No. TR-0059 (9250-03)-1, AF SAMSO-TR-70-301, August 1970.

114. Upp, J. W., Pepper, R. T., Kendall, E. G., and Rossi, R. C., High temperature properties of aluminum-graphite composites, Aerospace Report No. TR-0059 (6250-10)-9, AF SAMSO-TR-40-408, October 1970.

115. Pepper, R. T., Upp, J. W., Rossi, R. C., and Kendall, E. G., The tensile properties of a graphite-fiber reinforced Al-Si alloy, *Metallurgical Trans.*, 2, 117, 1971.

116. Alexander, J. A., Chuang, K. C., and Stuhrke, W. F., The elevated temperature activity in boron metal matrix composites, Advanced Fibrous Composites, 10th SAMPE Symp., San Diego, Calif., November 9 to 11, 1966.

117. Joseph, E., Myers, E. J., and Stuhrke, W. F., *J. Comp. Mater.*, 2, 56, 1968.

118. Shaw, C. W., Shepard, L. A., and Wulff, J., *Trans. ASM*, 57, 94, 1964.

119. Stuhrke, W. F., Solid state compatibility of boron aluminum composite material, Defense Metals Information Center Memorandum 243, May 1969, 43.

120. Clark, A. B., Effects of composite interfaces on mechanical properties, AFML-TR-68-43, 1967.

121. Hill, R. J. and Stuhrke, W. F., *Fiber Sci. Technol.*, 1 (1), 25, 1968.

122. Toth, I. J., An exploratory investigation of the time dependent mechanical behavior of composite materials, AFML-TR-69-9, April 1969.

123. Kreider, K., Dardi, L., and Prewo, K., Metal matrix composite technology, Technical Management Report for F33615-69-C-1539, AFML, Wright-Patterson AFB, Ohio, December 1969.

124. Antony, K. C. and Chang, W. H., *Trans. ASM*, 61, 550, 1968.

125. Getten, J. R. and Ebert, L. J., *Trans. ASM*, 62, 869, 1969.

126. Hecker, S. S., Hamilton, C. H., and Ebert, L. J., *Trans. ASM*, 62, 740, 1969.

127. Weeton, J. W. and Signorelli, R. A., Fiber-metal composite materials, NASA TN D-3530, Washington, D.C., August 1966.

128. Weeton, J. W. and Signorelli, R. A., Fiber reinforced metal composites under study at Lewis Research Center, in 14th Refractory Composites Meeting Proc., AFML-TR-68-129, May 1968, 309.

129. Petrasek, D. W. and Signorelli, R. A., Tungsten alloy fiber reinforced nickel base alloy composites for high temperature turbojet engine applications, in *Composite Materials: Testing and Design*, ASTM STP-460, American Society for Testing and Materials, 1969, 405.

130. McDanels, D. L., Jech, R. W., and Weeton, J. W., Metals reinforced with fibers, *Metals Progr.*, 78 (6), 118, 1960.

131. McDanels, D. L., Jech, R. W., and Weeton, J. W., Stress strain behavior of tungsten-fiber-reinforced copper composites, NASA TN D-1881, Washington, D. C., 1963.

132. Petrasek, D. W., Signorelli, R. A., and Weeton, J. W., Metallurgical and geometric factors affecting elevated-temperature tensile properties of discontinuous-fiber composites, NASA TN D-3886, Washington, D. C., March 1967.

133. Kelly, A. and Tyson, W. R., Fiber-strengthened materials, in *High Strength Materials*, 2nd Int. Materials Conf., Berkeley, Calif., Zackay, V. F., Ed., John Wiley & Sons, N.Y., 1965, 578.

134. Piehler, H. R., Plastic deformation and failure of silver-steel filamentary composite materials, TR 94-5, MIT Aeroelastic/Structures Research Laboratory, November 1963; (see also Piehler, H. R., *Trans. AIME*, 2333 (1), 12, 1965).

135. Karpinos, D. M., Umanskii, E. S., Rudenko, V. N., Tuchinskii, L. I., and Yu, G. D., Mechanical properties of nickel reinforced with tungsten fibers, translated from *Probl. Prochnosti*, 1, 78, 1969.

136. Ahmad, I. and Barranco, J. M., Strengthening of copper with tantalum (continuous) filaments, *Metallurgical Trans.*, 1, 989, 1970.

137. Reece, O. Y., TZM wire reinforced columbium composites, presented at 17th Refractory Composites Working Group Meeting, Williamsburg, Va., June 16 to 18, 1970.

138. Restall, J. E., Burwood-Smith, A., and Walles, K. F. A., The interaction between some reinforcing materials in Al and Ni-base matrices, *Metals and Materials*, 4 (11), 467, 1970.

139. Dean, A. V., The reinforcement of nickel-base alloys with high strength tungsten wires, *J. Inst. Met.*, 95, 79, 1967.

140. Blankenship, C. P., Oxide deformation and fiber reinforcement in a tungsten-metal oxide composite, NASA TN D-4475, Washington, D.C., March 1968.

141. McDanels, D. L. and Signorelli, R. A., Stress-rupture properties of tungsten wire from 1200° to 2500°F, NASA TN D-3467, Washington, D.C., July 1966.

142. **McDaniels, D. L., Signorelli, R. A., and Weeton, J. W.,** Analysis of stress-rupture and creep properties of tungsten-fiber-reinforced copper composites, NASA TN D-4173, Washington, D.C., September 1967.
143. **Petrasek, D. W., Signorelli, R. A., and Weeton, J. W.,** Refractory metal fiber nickel alloy composites for use at high temperatures, presented at 12th National Symp., SAMPE, Anaheim, Calif., October 1967, and NASA TM X-52342, Washington, D.C., 1967.
144. **Weeton, J. W. and Signorelli, R. A.,** Fiber reinforced superalloys and heat resistant matrix composites, presented at AIME Meeting Composites: State-of-the-Art Session, Detroit, Mich., October 1971.
145. **Klein, M. J., Domes, R. B., and Metcalfe, A. G.,** Tungsten fiber reinforced oxidation resistant columbium alloys, Report on Navy Contract N00019-71-C-0158, May 31, 1971, available through I. Machlin, Naval Air Systems Command, Washington, D.C. 20360, AIR-520312.
146. **Brennan, J. J.,** High temperature metal matrix composites, presented at 17th Refractory Composites Working Group Meeting, Williamsburg, Va., June 16 to 18, 1970.
147. **McDaniels, D. M.,** Electrical resistivity and conductivity of tungsten-fiber-reinforced copper composites, NASA TN D-3590, Washington, D.C., August 1966.
148. **Raw, P. M.,** Fibre-reinforced copper of high strength and high electrical conductivity, *Appl. Mat. Res.*, 4 (3), 176, 1965.
149. **Parikh, N. W.,** Fiber-reinforced metals and alloys, Rep. No. ARL 2193-6, Armour Research Foundation, Illinois Institute of Technology, March 22, 1961.
150. **Stevenson, W. D., Jr.,** *Elements of Power System Analysis*, McGraw-Hill Book Co., Inc., 1955, 25, 355.
151. **Brenner, S. S.,** *J. Appl. Phys.*, 33, 33, 1962.
152. **Brenner, S. S.,** *J. Metals*, 14 (11), 808, 1962.
153. **Hahn, H. and Kershaw, J. P.,** Development of ceramic fiber reinforced metals, presented at the 2nd Interamerican Conf. on Materials Technology, Mexico City, Mexico, August 24 to 27, 1970.
154. **Sutton, W. H. and Chorne, J.,** Potential of oxide fiber reinforced metals, ASM Seminar Volume 1964, Ch. 9, ASM, Metals Park, Ohio, 1965, 173.
155. **Sutton, W. H. and Chorne, J.,** *Metals Engin. Quart.*, 3 (1), 44, 1963.
156. **Mehan, R. L.,** Fabrication and evaluation of sapphire whisker reinforced aluminum composites, *Metal Matrix Composites*, American Society for Testing Materials, ASTM STP-438, 1968, 29.
157. **Mehan, R. L. and Jakas, R.,** Behavior study of sapphire wool aluminum and aluminum alloy composites, AFML-TR-69-62, January 1969.
158. **Mehan, R. L.,** *J. Comp. Mater.*, 4, 90, 1970.
159. **Divecha, A., Lare, P., and Hahn, H.,** Silicon carbide whisker-metal matrix composites, AFML-TR-69-7, May 1969.
160. **Herzog, J. A.,** Methods and devices for mechanical tests on filaments and whiskers, AFML-TR-66-417, AFML, February 1967.
161. **Herzog, J. A.,** The metal composites, their reinforcing components, and their problem areas, AFML-TR-67-50, AFML, March 1967.
162. **Herzog, J. A.,** Filaments, fibers, and metal matrix composites, AFML-TR-67-244, October 1967.
163. **Herzog, J. A.,** Development and potential of whiskers for composites, AFML-TR-67-326, December 1967.
164. **Grenier, P. and Marchal, M.,** *Mem. Sci. Rev. Metallurg.*, 4, 345, 1968.
165. **Hahn, H.,** Materials Technology — An Interamerican Approach, *Amer. Soc. Mech. Eng.*, New York, 1968, 231.
166. **Baker, A. A. and Cratchley, D.,** *Appl. Mat. Res.*, 5, 92, 1966.
167. **Parratt, N. J.,** *Chem. Eng. Progr.*, 62, 61, 1966.
168. **Halpin, J. C.,** *J. Comp. Mater.*, 3, 732, 1969.
169. **Hill, W. H. and Shimmin, K. D.,** Elevated temperature dynamic elastic moduli of various metallic materials, WADD Tech. Report 60, 438, 1961.
170. **Kelly, A. and Tyson, W. R.,** *J. Mech. Phys. Solids*, 14, 177, 1966.
171. **Ordway, F., Lare, P. J., and Hermann, R. A.,** Silicon carbide whisker-metal matrix composites, AFML-TR-71-252, ARTECH Corporation, October 1971, Final Report, Contract No. F33615-69-C-1187.
172. **Lare, P. J., Ordway, F., and Hahn, H.,** Research on whisker-reinforced metal composites, (U), Naval Air Systems Command, Contract N00019-70-C-0204, Final Report, December 1971.
173. **Young, J. H.,** Advanced composite material structural hardware development and testing program, AFML Report TM 69-249, G. E. April 30, 1969.
174. **Kelly, A. and Davies, G. J.,** *Metallurgical Rev.*, 10, 1, 1965.
175. **Rosen, B. W.,** The strength and stiffness of fibrous composites, in *Modern Composite Materials*, 1967, Ch. 3.
176. **Ebert, L. J., Fedor, R. J., Hamilton, C. H., Hecker, S. S., and Wright, P. K.,** The analytical approach to composite behavior, AFML-TR-69-129, June 1969.
177. **Schaefer, W. H. and Christian, J. L.,** Evaluation of the structural behavior of filament reinforced metal matrix composites, AFML-TR-69-36, Vol. III, January 1969.
178. **Toth, I. J.,** An exploratory investigation of time dependent mechanical behavior of composite materials, TRW Materials Technology Report ER 7274-2, September 1969.
179. **Schaefer, W. H. and Christian, J. L.,** Evaluation of the structural behavior of filament reinforced metal matrix composites, AFML-TR-69-36, Vol. II, January 1969.

180. Klein, M. J., Reid, M. L., and Metcalfe, A. G., Compatibility studies for viable titanium matrix composites, Solar Quarterly Report, October 1969 to December 1969, AF Contract F33615-68-C-1423.
181. Forest, J. D. and Christian, J. L., Development and application of high-matrix strength aluminum-boron, Metal Matrix Composites Session of the Materials Engineering Congress, October 16, 1969.
182. Cooper, G. A., *J. Mech. Phys. Solids,* 14, 103, 1966.
183. Stowell, E. Z. and Liu, T. S., *J. Mech. Phys. Solids,* 9, 242, 1961.
184. Jackson, P. W. and Cratchley, D., *J. Mech. Phys. Solids,* 14, 49, 1966.
185. Kreider, K. G., Mechanical testing of metal matrix composites, *Composite Materials: Testing and Design,* American Society for Testing and Materials Special Technical Publication 460, 1969.
186. Toth, I. J., An exploratory investigation of the time dependent mechanical behavior of composite materials, AFML-TR-69-9, April 1969.
187. Kreider, K. G., Dardi, L., and Prewo, K. M., Progress Report, Air Force Contract F33615-69-C-1539, January 1, 1970, February 28, 1970.
188. Adsit, N. R. and Forest, J. D., Compression testing of aluminum-boron composites, *Composite Materials; Testing and Design,* American Society for Testing and Materials Special Technical Publication 460, 1969.
189. Toth, I. J., Creep and fatigue behavior of unidirectional and cross-plied composites, *Composite Materials: Testing and Design,* American Society for Testing and Materials Special Technical Publication 460, 1969.
190. Antony, K. C. and Chang, W. H., *Trans. ASM,* 61, 550, 1968.
191. Veltri, R. and Galasso, F., *Nature,* 220, 781, 1968.
192. Compton, W. A., Stewart, K. P., and Monew, H., Composite materials for turbine compressors, AFML-TR-68-31, June 1968.
193. DeSilva, A. R. T., *J. Mech. Phys. Solids,* 16, 169, 1968.
194. Kelly, A. and Tyson, W. R., *J. Mech. Phys. Solids,* 14, 177, 1966.
195. Ellison, E. G. and Boone, D. H., *J. Less-Common Metals,* 13, 103, 1967.
196. Metcalfe, A. G. and Rose, F. K., Testing of thin gage materials, AFML-TR-68-64, 1968.
197. Galasso, F., Salkin, M., Kuehl, D., and Patarini, U. A., *Trans. TMS-AIME,* 236, 1748, 1966.
198. Kraft, R. W., *J. Metals,* 18, 192, 1966.
199. Salkind, M., Lemkey, F., George, F., and Bayles, B. J., Eutectic composites by unidirectional solidification, *Advanced Fibrous Reinforced Composites,* Vol. 10, SAMPE, Western Periodicals Co., North Hollywood, California, 1966, F35, 44.
200. Salkind, M., Bayles, B. J., George, F., and Tice, W., Investigation of Fracture Mechanisms, Thermal Stability and Hot Strength of Controlled Polyphase Alloys, Final Report Contract NOw 64-0433-d, April, 1965. (See also Bayles, B. J., Ford, J. A., and Salkind, M. J., *Trans. AIME,* 239, 844, 1967.)
201. Thompson, E. R. and George, F. D., Investigation of the Structure and Properties of the Ni$_3$Al-Ni$_3$Nb Eutectic Alloy, Final Report Contract N00019-69-C-1062, July 31, 1969.
202. Weiss, H. and Wilhelm, M., *Z. Phys.,* 176, 399, 1963.
203. Weiss, H., *Met. Trans.,* 2, 1513, 1971.
204. Mollard, F. R. and Flemings, M. C., *Trans. AIME,* 239, 1526, 1967.
205. Mollard, F. R. and Flemings, M. C., *Trans. AIME,* 239, 1534, 1967.
206. Salkind, M., George, F., Lemkey, F., and Bayles, B., Investigation of the Creep, Fatigue, and Transverse Properties of Al$_3$Ni Whisker and CuAl$_2$ Platelet Reinforced Aluminum, Final Report Contract NOw 65-0384d, May 1966.
207. George, F. D., Ford, J. A., and Salkind, M. J., The effect of fiber orientation and morphology on the tensile behavior of Al$_3$Ni whisker reinforced aluminum, *Metal Matrix Composites,* ASTM STP 438, 1968, 59.
208. Chadwick, G., *J. Inst. Metals,* 91, 169, 1963.
209. Graham, L. and Kraft, R., *Trans. AIME,* 236, 94, 1966.
210. Salkind, J., Interfacial stability of eutectic composites, in *Interfaces in Composites,* ASTM STP STP 452, 149.
211. Hunt, J. D. and Jackson, K. A., *Trans. AIME,* 236, 843, 1966.
212. Yue, A. S., *Met. Trans.,* 1, 19, 1970.
213. Verhoeven, J. D. and Homer, R. H., *Met. Trans.,* 1, 3437, 1970.
214. Jackson, K. A., and Hunt, J. D., *Trans. AIME,* 236, 1129, 1966.
215. Cline, H. E., *Trans. AIME,* 245, 2205, 1969.
216. Hopkins, R. H. and Kraft, R. W., *Trans. AIME,* 242, 1627, 1968.
217. Cline, H. E., *Trans. AIME,* 239, 1489, 1967.
218. Jackson, K. A., *Trans. AIME,* 242, 1275, 1968.
219. Bertorello, H. R. and Biloni, H., *Trans. AIME,* 245, 1373, 1969.
220. Walter, J. L., Cline, H. E., and Koch, E. F., *Trans. AIME,* 245, 2073, 1969.
221. Jaffrey, D. and Chadwick, G. A., *Trans. AIME,* 245, 2435, 1969.
222. Jaffrey, D. and Chadwick, G. A., *Met. Trans.,* 1, 3389, 1970.
223. Cline, H. E., Walter, J. L., Lifshin, E., and Russell, R. R., *Met. Trans.,* 2, 189, 1971.
224. Yue, A. S. and Crossman, F. W., *Met. Trans.,* 1, 322, 1970.
225. Beghi, G., Piatti, G., and Street, K. N., *J. Mater. Sci.,* 6, 118, 1971.
226. Cline, H. E., *Acta Met.,* 19, 481, 1971.

227. Cline, H. E., Walter, J. L., Koch, E. F., and Osika, L. M., *Acta Met.*, 19, 405, 1971.
228. Copley, S. M. and Kear, B. H., Temperature and orientation dependence of the flow stress in off-stoichiometric Ni₃Al (a' Phase), *Trans. AIME*, 239, 977, 1967.
229. Guard, R. W. and Westbrook, J. H., Alloying behavior of Ni₃Al (a' phase), *Trans. AIME*, 215, 807, 1959.
230. Thompson, E. R. and Lemkey, F. D., *Trans. ASM*, 62, 140, 1969.
231. Savitskiy, Y. M., Kopetskiy C. V., and Arskaya, Y. P., The effect of temperature on the plastic deformation and mechanical properties of some intermetallic compounds, *Russ. Met.*, 6, 85, 1964.
232. Kornilov, I. I., Shinayev, A. Y., and Pylayev, Y. N., Creep of some intermetallic compounds, *Russ. Met.*, 5, 67, 1963.
233. Mints, R. S., Belyaev, G. F., and Malkov, Y. S., Investigation of the high-temperature strength of Ni₃Al-Ni₃Nb alloys, *Russ. Met.*, 4, 15, 1963.
234. Mints, R. S., Belyaev, G. F., and Malkov, Y. S., Equilibrium diagram of the Ni₃Al-Ni₃Nb system, *Russ. J. Inorg. Chem.*, 7 (10), 1236, 1962.
235. Salkind, M. J., George, F. D., Bayles, B. J., and Ford, J. A., *Chem. Eng. Progr.*, 62, 52, 1966.
236. Rudy, E., Ternary Phase Equilibria in Transition Metal-Boron-Carbon Silicon Systems, Part V, Compendium of Phase Diagram Data, AFML TR-65-2, June 1969.
237. Stover, E. R. and Wulff, J., *Trans. AIME-TMS*, 215, 127, 1959.
238. Edwards, R. and Raine, T., Plansee Proceedings, 1952, 232.
239. Bibring, H. and Seibel, G., *C. R. Acad. Sci.* (Paris), 268, Series C., 144, 1969.
240. Lemkey, F. D. and Thompson, E. R., *Met. Trans.*, 2, 1537, 1971.
241. Thompson, E. R. and Lemkey, F. D., *Met. Trans.*, 1, 2799, 1970.
242. B-1900 Technical Data, Austenal Microcast Division, Howmet Corp., 1965.
243. Hoover, W. R. and Hertzberg, R. W., *Met Trans.*, 2, 1282, 1971.
244. Hoover, W. R. and Hertzberg, R. W., *Met. Trans.*, 2, 1289, 1971.
245. Salkind, M. J. and George, F. D., *Trans. AIME*, 242, 1237, 1968.
246. Salkind, M., George, F., and Tice, W., *Trans. AIME*, 245, 2339, 1969.
247. Hertzberg, R. W., Lemkey, F. D., and Ford, J. A., *Trans. AIME*, 233, 342, 1966.
248. Bibring, H., Rabinovitch, M., and Seibel, G., *C.R. Acad. Sci.* (Paris), 268, 1666, 1969.
249. Redden, T. and Barker, J., *Met. Progr.*, 87, 107, 1965.
250. Thompson, E. R., Koss, D. A., and Chesnutt, J. C., *Met. Trans.*, 1, 2807, 1970.
251. Koss, D. A. and Copley, S. M., *Met. Trans.*, 2, 1557, 1971.
252. Thompson, E. R., *J. Comp. Mater.*, 5, 235, 1971.
253. Pattnaik, A. and Lawley, A., *Met. Trans.*, 2, 1529, 1971.
254. Kossowsky, R., Johnston, W. C., and Shaw, B. J., *Trans. AIME*, 245, 1219, 1969.
255. Crossman, F. W. and Yue, A. S., *Met. Trans.*, 2, 1545, 1971.
256. Colling, D. A. and Kossowsky, R., *Met. Trans.*, 2, 1523, 1971.
257. Sendeckyj, G. P., On strength of composite materials, ASTM-AIME Meeting, Anaheim, California, April 20, 1971.
258. Pagano, N. J. and Tsai, S. W., *Composite Materials Workshop*, Technomic Pub. Co. Inc., Stamford, Connecticut, 1968, 1.
259. Ashton, J. E., Halpin, J. C., and Petit, P. H., *Primer on Composite Materials: Analysis*, Technomic Pub. Co. Inc., Stamford, Connecticut, 1969, 124.
260. Halpin, J. C., Pagano, N. J., Whitney, J. M., and Wu, E. M., Characterization of anisotropic composite materials, *Composite Materials: Testing and Design*, ASTM STP 460, 1969, 37.
261. Whitney, J. M. and Pagano, N. J., Design and fabrication of tublar specimens for composite characterization, ASTM-AIME Meeting, Anaheim, California, April 20, 1971.
262. Lynch, C. T., Kershaw, J. P., and Collins, B. R., *CRC Crit. Rev. Solid State Sci.*, 1, (4), 481, 1970.
263. Hill, R., *J. Mech. Phys. Solids*, 11, 357, 1963.
264. Zweben, C., Tensile strength of fiber-reinforced composites: basic concepts and recent developments, *Composite Materials: Testing and Design*, ASTM STP 460, ASTM, 1969, 528.
265. Rosen, B. W., Mechanics of Composite Strengthening, in *Fiber Composite Materials*, American Society for Metals, Metals Park, Ohio, 1965.
266. Structural Design Guide for Advanced Composite Applications, Air Force Materials Laboratory Publication, 1st ed., August, 1969.
267. Hermans, J. J., Proc. Konig, Nederl., Akad van Weteschappen, Amsterdam, B70, (1)1, 1967.
268. Eshelby, J. D., Elastic inclusions and inhomogeneities, in *Progress in Solid Mechanics*, Vol. 2, Sneddon and Hill, Eds., North Holland, Amsterdam, 1961, 89.
269. Hill, R., *J. Mech. Phys. Solids*, 13, 189, 1965.
270. Kilchinskii, A. A., *Prikl. Mekh.*, 1 (12), 65, 1965 (Russian).
271. Whitney, J. M. and Reily, M. B., *J. AIAA*, 4, 1537, 1966.
272. Sokolnikof, I. S., *Mathematical Theory of Elasticity*, McGraw-Hill, New York, 1956, Ch. 7.
273. Paul, B., *Trans. AIME*, 218, 36, 1960.
274. Hashin, Z. and Rosen, B. W., *J. Appl. Mech., Trans. ASME*, 31, 223, 1964 (Errata, *J. Appl. Mech.*, 32, 219, 1965).

275. Hashin, Z., *J. Mech. Phys. Solids*, 13, 119, 1965.
276. Van Fo Fy, G. A., *Prikl. Mekh.*, 1 (2), 110, 1965 (Russian).
277. Herman, L. R. and Pister, K. S., Composite properties of filament-resin systems, ASME P.N. 63, WA-239, paper presented at the ASME Annual Meeting, Philadelphia, November 17 to 22, 1963.
278. Wilson, H. B., Jr. and Hill, J. L., Mathematical Studies of Composite Materials, Rohm and Haas Special Report No. S-50 AD 468 569, 1965.
279. Adams, D. F. and Bloom, J. M., Nonconservative Behavior of Composite Materials, Quarterly Progress Report No. 3, AFML Contract AF33(615)-5198, 1967.
280. Adams, D. F. and Doner, D. R., *J. Comp. Mater.*, 1, 4, 1967.
281. Adams, D. F. and Doner, D. R., *J. Comp. Mater.*, 1, 152, 1967.
282. Adams, D. F., Doner, D. R., and Thomas, R. L., Mechanical behavior of fiber-reinforced composite materials, AFML-TR-67-96, 1967.
283. Pickett, G., Analytical Procedures for Predicting the Mechanical Properties of Fiber Reinforced Composites, AFML-TR-65-220.
284. Pickett, G. and Johnson, M. N., Analytical Procedures for Predicting the Mechanical Properties of Fiber Reinforced Composites, AFML-TR-65-222, pt. 2.
285. Bloom, J. M. and Wilson, H. B., *J. Comp. Mater.*, 1, 268, 1967.
286. Foye, R. L., An evaluation of various engineering estimates of the transverse properties of unidirectional composites, SAMPE, 10, G-31, 1966.
287. Foye, R. L., Structural Composites, Quarterly Progress Reports Nos. 1 and 2, AFML Contract No. AF33(615)-5150, 1966.
288. Tsai, S. W., *Some Fundamental Principles Associated with Composite Materials*, The Plastics Institute Transactions and Journal, Pergamon Press Ltd., 1969, 391.
289. Halpin, J. C. and Tsai, S. W., Environmental Factors in Composite Materials Design, AFML-TR-67-423, 1967.
290. Tsai, S. W., Mechanics of Composite Materials, part II, AFML-TR-66-149, November 1966.
291. Lekhnitskii, S. G., *Anisotropic Plates*, 2nd ed., OGIZ, Moscow-Leningrad, 1947. Translated by S. W. Tsai and T. Cheron, Gordon and Breach, 1968.
292. Ashton, J. E., *J. Comp. Mater.*, 4, 162, 1970.
293. Whitney, J. M., *J. Comp.*, 4, 192, 1970.
294. Taylor, R. J. et al., Mechanical behavior of aluminum-boron composites, AFML-TR-68-385, August 1969.
295. Ebert, L. J., Hamilton, C. H., and Hecker, S. S., Analytical approach to composite behavior, AFML-TR-68-71, March 1968.
296. Haener, J. and Ashbaugh, N., *J. Comp. Mater.*, 1, 54, 1967.
297. Taylor, R. J., Dolowy, J. F., Jr., and Shimizu, H., Application of composite materials to ramjet inlet structures, AFML-TR-68-85, Vol. I, May 1968.
298. Foye, R. L., Advanced design concepts for advanced composite airframes, AFML-TR-68-91, Vol. I, July 1968.
299. Norris, C. B. and McKinnon, P. F., Compression, tension and shear tests on pellow-poplar plywood panels of sizes that do not buckle with tests made at various angles to the face grain, Forest Products Laboratory Report No. 1328, Great Britain, 1946.
300. Norris, C. B., Strength of orthotropic materials subjected to combined stresses, Forest Products Laboratory Report No. 1816, Great Britain, 1950.
301. Hill, R., A theory of the yielding and plastic flow of anisotropic metals, *Proc. R. Soc., Ser. A*, 193, 1948.
302. Kaminski, B. E. and Lantz, R. B., Strength theories of failure for anisotropic materials, *Composite Materials: Testing and Design*, ASTM STP 460, American Society for Testing and Materials, 1969.
303. Jenkin, C. F., Report on materials of construction used in aircraft and aircraft engines, Great Britain Aeronautical Research Committee, 1920.
304. Marin, J. J., *Aeronautical Sci.*, 24(3), 187, 1957.
305. Greszczuk, L. B., Elastic constants and analysis methods for filament wound shell structures, Douglas Aircraft Company Report No. SM 45849, Appendix A, 1964.
306. Azzi, V. D. and Tsai, S. W., *Exp. Mech.*, 5, 283, 1965.
307. Hoffman, D., *J. Comp. Mater.*, 1, 200, 1967.
308. Hu, L. W., *J. Franklin Inst.*, 265 (3), 1958.
309. Waddoups, M. E., Advanced composite materials mechanics for the design and stress analyst, General Dynamics Division Report FZM-4763, 1967.
310. Tsai, S. W. and Wu, E. M., *J. Comp. Mater.*, 5, 58, 1971.
311. Hecker, S. S., Hamilton, C. H., and Ebert, L. J., *J. Mater.*, 5, 868, 1970.
312. Tetleman, A. S., Fracture processes in fiber composite materials, *Composite Materials: Testing and Design*, ASTM STP 460, American Society for Testing and Materials, 1969, 473.
313. Tetleman, A. S. and McEvily, A. J., *Fracture of Structural Materials*, John Wiley & Sons, New York, 1967.
314. Kreider, K. G., Dardi, L., and Prewo, K., Metal matrix composite technology, AFML-TR-71-204, December 1971.
315. Toth, I., Research for elevated temperature mechanical properties of B/Ti composites, Report on Contract F33615-71-C-1044, July 1971.

316. **Miller, M. F., Schaefer, W. H., and Weisinger, M. D., et al.,** Development of improved metal-matrix fabrication techniques for aircraft, AFML-TR-71-181, July 1971.

317. **Happe, R. A. and Yeast, A. J.,** Evaluation of boron-aluminum composite material for space structures, 15th National SAMPE Symposium, Los Angeles, California, April 1969.

318. **Hersh, M. S. and Duffy, E. R.,** Development of fabrication methods for aluminum-boron composite aircraft structures, DMIC Memorandum 243, 112-124, May 1969.

319. **Kutner, G. L.,** Metal-matrix composites for propeller blade applications, DMIC Memorandum 243, 107-111, May 1969.

320. **Young, J. H. and Carlson, R. G.,** Advanced composite material structural hardware development and testing program, AFML-TR-70-140, Vol. I, July 1970.

321. **Alexander, J. A.,** Review of metal composites programs, 17th Refractory Working Group Meeting, Williamsburg, Virginia, June 15, 1970.

322. **Toth, I. J., Brentnall, W. D., and Menke, G. D.,** Aluminum matrix composites, AIME Fall Meeting, Detroit, Michigan, October 20, 1971.

Printed and bound by CPI Group (UK) Ltd, Croydon, CR0 4YY

22/10/2024

01777600-0018